Probiotic Bacteria and Their Effect on Human Health and Well-Being

World Review of Nutrition and Dietetics

Vol. 107

Series Editor

Berthold Koletzko Munich

Probiotic Bacteria and Their Effect on Human Health and Well-Being

Volume Editors

Alfredo Guarino Naples
Eamonn M.M. Quigley Cork
W. Allan Walker Boston, Mass.

22 figures, and 17 tables, 2013

KARGER

Basel · Freiburg · Paris · London · New York · New Delhi · Bangkok ·
Beijing · Tokyo · Kuala Lumpur · Singapore · Sydney

Alfredo Guarino
Department of Translational Medical Science-
Section of Pediatrics
University of Naples Federico II
Naples, Italy

Eamonn M.M. Quigley
Department of Medicine
University College Cork
Cork, Ireland

W. Allan Walker
Department of Pediatrics
Harvard Medical School
Boston, Mass., USA

Library of Congress Cataloging-in-Publication Data

Probiotic bacteria and their effect on human health and well-being / volume
editors, Alfredo Guarino, Eamonn M.M. Quigley, W. Allan Walker.
      p. ; cm. -- (World review of nutrition and dietetics, ISSN 0084-2230
; v. 107)
  Includes bibliographical references and index.
  ISBN 978-3-318-02324-4 (hard cover : alk. paper) -- ISBN 978-3-318-02325-1
(e-ISBN)
  I. Guarino, Alfredo, 1955- II. Quigley, Eamonn M. M. III. Walker, W.
Allan. IV. Series: World review of nutrition and dietetics ; v. 107.
0084-2230
  [DNLM: 1. Probiotics--therapeutic use. 2. Gastrointestinal
Diseases--microbiology. 3. Gastrointestinal Tract--microbiology. 4.
Metagenome--physiology. 5. Probiotics--adverse effects. W1 WO898 v.107
2013 / QU 145.5]
  RM666.P835
  615.3'29--dc23
                                                    2013006924

Bibliographic Indices. This publication is listed in bibliographic services, including Current Contents® and PubMed/MEDLINE.

© Copyright 2013 by S. Karger AG, P.O. Box, CH–4009 Basel (Switzerland)
www.karger.com
Printed in Germany on acid-free and non-aging paper (ISO 9706) by Kraft Druck, Ettlingen
ISSN 0084–2230
e-ISSN 1662–3975
ISBN 978–3–318–02324–4
e-ISBN 978–3–318–02325–1

Contents

List of Contributors

Louis M.A. Akkermans
Gastrointestinal Physiology, Utrecht University
University Medical Center Utrecht
Amsterdam University College
Gezichtslaan 27
NL-3723 GB Bilthoven (The Netherlands)

Eoin Barrett
Alimentary Pharmabiotic Centre
Biosciences Institute
Department of Microbiology
University College Cork
Cork (Ireland)

Patrizia Brigidi
Department of Pharmaceutical Science
University of Bologna
Via Belmeloro 6
IT-40126 Bologna (Italy)

Eugenia Bruzzese
Department of Translational Medical Science-
Section of Pediatrics
University of Naples 'Federico II'
Via Sergio Pansini 5
IT-80131 Naples (Italy)

Vittoria Buccigrossi
Department of Translational Medical Science-
Section of Pediatrics
University of Naples 'Federico II'
Via Sergio Pansini 5
IT-80131 Naples (Italy)

Roberto Berni Canani
Food Allergy Unit
Department of Translational Medicine – Pediatric
Section, University of Naples 'Federico II'
Via Sergio Pansini 5
IT-80131 Naples (Italy)

Maria C. Collado
Instituto de Agroquímica y Tecnología de
Alimentos (IATA-CSIC)
ES-46980 Valencia (Spain)

Linda Cosenza
Department of Translational Medical Science-
Section of Pediatrics
University of Naples 'Federico II'
Via Sergio Pansini 5
IT-80131 Naples (Italy)

Margherita Di Costanzo
Department of Translational Medical Science-
Section of Pediatrics
University of Naples 'Federico II'
Via Sergio Pansini 5
IT-80131 Naples (Italy)

John F. Cryan
Alimentary Pharmabiotic Centre
Biosciences Institute
Cork (Ireland)

Timothy G. Dinan
Alimentary Pharmabiotic Centre
Biosciences Institute
Cork (Ireland)

Gerald F. Fitzgerald
Department of Microbiology
4th Floor
Food Science & Technology Building
University College Cork
Cork (Ireland)

Sofia D. Forssten
Danisco Sweeteners
Active Nutrition
DuPont Nutrition and Health
Sokeritehtaantie 20
FI-02460 Kantvik (Finland)

Viviana Granata
Department of Translational Medical Science-
Section of Pediatrics
University of Naples 'Federico II'
Via Sergio Pansini 5
IT-80131 Naples (Italy)

Stefano Guandalini
Section of Pediatric Gastroenterology
University of Chicago
MC 4065
5841 S. Maryland Ave.
Chicago, IL 60637 (USA)

Alfredo Guarino
Department of Translational Medical Science-
Section of Pediatrics
University of Naples 'Federico II'
Via Sergio Pansini 5
IT-80131 Naples (Italy)

Francisco Guarner Aguilar
Digestive Diseases Research Unit
University Hospital Vall d'Hebron
Paseo Vall d'Hebron 118
ES-08035 Barcelona (Spain)

Colin Hill
Alimentary Pharmabiotic Centre
Department of Microbiology
University College Cork
Cork (Ireland)

Iva Hojsak
Referral Center for Pediatric Gastroenterology
and Nutrition
Children's Hospital Zagreb
University of Zagreb School of Medicine
Klaićeva 16
HR–10000 Zagreb (Croatia)

Fandi Ibrahim
School of Science, Technology, and Health
University Campus Suffolk
Waterfront Building
Neptune Quay
Ipswich IP4 1QJ (UK)

Flavia Indrio
Department of Pediatrics
University of Bari
Piazza Giulio Cesare
IT-70124 Bari (Italy)

Erika Isolauri
Institute of Clinical Medicine
Department of Paediatrics
Turku University Hospital
FIN-20520 Turku (Finland)

Koichi S. Kobayashi
Department of Microbial and Molecular
Pathogenesis
College of Medicine
Texas A&M Health Science Center
415A Reynolds Medical Building
College Station, TX 77843-1114 (USA)

Kirsi Laitinen
Institute of Biomedicine
Functional Foods Forum
University of Turku
FIN-20014 Turku (Finland)

Ludovica Leone
Department of Translational Medical Science-
Section of Pediatrics
University of Naples 'Federico II'
Via Sergio Pansini 5
IT-80131 Naples (Italy)

Andrea Lo Vecchio
Department of Translational Medical Science-
Section of Pediatrics
University of Naples 'Federico II'
Via Sergio Pansini 5
IT-80131 Naples (Italy)

Raakel Luoto
Department of Paediatrics
Satakunta Central Hospital
Sairaalantie 3
FIN-28500 Pori (Finland)

Susan V. Lynch
Division of Gastroenterology
Department of Medicine
University of California
513 Parnassus Ave.
San Francisco, CA 94143-0538 (USA)

Lorenzo Morelli
Preside/Dean Facoltà di Agraria
UCSC Piacenza e Cremona
Via Emilia Parmense 84
IT-29122 Piacenza (Italy)

Linda Mulder
Winclove
Hulstweg 11
NL-1032 LB Amsterdam (The Netherlands)

Eileen F. Murphy
Alimentary Health Ltd.
Building 2800
Cork Airport Business Park
Kinsale Road
Cork (Ireland)

Josef Neu
Neonatal Biochemical Nutrition and GI
Development Laboratory
Department of Pediatrics
Division of Neonatology
University of Florida Box J296
1600 SW Archer Road
Room HD 513
Gainesville, FL 32610-0296 (USA)

Emanuele Nicastro
Department of Translational Medical Science
Section of Pediatrics
University of Naples 'Federico II'
Via Sergio Pansini 5
IT-80131 Naples (Italy)

Rita Nocerino
Department of Translational Medical Science-
Section of Pediatrics
University of Naples 'Federico II'
Via Sergio Pansini 5
IT-80131 Naples (Italy)

Paul W. O'Toole
Department of Microbiology
4th Floor
Food Science & Technology Building
University College Cork
Cork (Ireland)

Arthur C. Ouwehand
Danisco Sweeteners
Active Nutrition
DuPont Nutrition and Health
Sokeritehtaantie 20
FIN-02460 Kantvik (Finland)

Tiffany J. Patton
Section of Pediatric Gastroenterology
University of Chicago
MC 4065,
5841 S. Maryland Ave.
Chicago, IL 60637 (USA)

Vincenza Pezzella
Department of Translational Medical Science-
Section of Pediatrics
University of Naples 'Federico II'
Via Sergio Pansini 5
IT-80131 Naples (Italy)

Bénédicte Pigneur
Université Paris Descartes
Sorbonne Paris Cité
FR-75006 Paris (France)

Susan E. Power
Department of Microbiology
Alimentary Pharmabiotic Centre
5th Floor (Lab 5.27)
Biosciences Institute
University College Cork
Cork (Ireland)

Eamonn M.M. Quigley
Alimentary Pharmabiotic Centre
Department of Medicine
University College Cork
Cork (Ireland)

Giuseppe Riezzo
Laboratory of Experimental Pathophysiology
IRCCS 'S de Bellis'
IT-70013 Castellana Grotte (Italy)

Ger T. Rijkers
Department of Sciences
Roosevelt Academy
Lange Noordstraat 1
PO Box 94
NL-4330 AB Middelburg (The Netherlands)

Virginia Robles Alonso
Digestive Diseases Research Unit
University Hospital Vall d'Hebron
Paseo Vall d'Hebron 118
ES-08035 Barcelona (Spain)

Frans M. Rombouts
Laboratory of Food Microbiology
Department of Agrotechnology and Food
Sciences
Wageningen University
Bomenweg 2
NL-6703 HD Wageningen (The Netherlands)

R. Paul Ross
Teagasc Food Research Centre,
Moorepark
Fermoy, Co. Cork
(Ireland)

Henna Röytiö
University of Turku
Functional Foods Forum,
FI-20014 Turun Yliopisto (Finland)

Eliana Ruberto
Department of Translational Medical Science-
Section of Pediatrics
University of Naples 'Federico II'
Via Sergio Pansini 5
IT-80131 Naples (Italy)

Frank M. Ruemmele
Assistance Publique-Hôpitaux de Paris
Hôpital Necker-Enfants Malades
Service de Gastroentérologie pédiatrique
149 rue de Sèvres
FR-75015 Paris (France)

Seppo Salminen
Functional Foods Forum
University of Turku
FIN-20014 Turku (Finland)

Raanan Shamir
Sackler Faculty of Medicine
Tel Aviv University
IL-69978 Tel Aviv (Israel)

Fergus Shanahan
Alimentary Pharmabiotic Centre
Biosciences Institute
University College Cork
Cork (Ireland)

Catherine Stanton
Teagasc Food Research Centre
Moorepark
Fermoy, Co. Cork
(Ireland)

W. Allan Walker
Department of Nutrition
Harvard School of Public Health
655 Huntington Avenue
Building II 3rd Floor
Boston, MA 02115 (USA)

Preface

Probiotic Bacteria and Their Effect on Human Health and Well-Being provides an update on probiotics which is directed at physicians, biologists biotechnologists, and researchers working in the food industry, agriculture, and the environmental, basic sciences and in health care.

Each human inhabitant of the planet belongs to 1 of 3 distinct enterotypes. Enterotypes are defined by the predominant bacterial phyla located in the intestine. Therefore, our microflora may be regarded as an individual personal feature like a blood group, providing a distinct tag to individuals. However, our intestinal microbiome is strongly affected by genetic, nutritional, and other external factors. The microbiome evolves with age and is modified by nutritional habits. Children on a fiber-rich diet in Africa harbor an intestinal microflora different from their age-matched peers in Florence (Italy) who eat the typical Western style diet of a wealthy European country. Whether the different microbial patterns have an effect on health is not conclusively known, but appears very likely. Several recent papers have described specific changes of intestinal microflora in association with inflammatory bowel diseases, atopy, intestinal functional disorders, and obesity. The pattern of microflora aberrations is often age-specific or condition-specific.

These findings are important for our understanding of the pathophysiology and risk factors of human diseases. The structure of the intestinal microflora may be exploited for practical diagnostic purposes, and thus several titles of papers describing the microflora composition in several diseases use the word 'signature'. The concept of a 'microbial signature' of a given disease indicates a role of a specific pattern of microflora with that disease although there is often no direct evidence of a cause-effect link. The novel concept is that we may use specific microbial tags as biomarkers of a disease, to diagnose it, to monitor its evolution, and eventually to predict its response to treatment.

This scenario opens the opportunity for targeting the intestinal microflora with the use of probiotics. Clinical indications for probiotics include prevention and treatment of an increasing number of conditions. Probiotics are used as drugs, usually in lyophilized preparations, or in addition to foods, as additives, modifiers, or functional foods. While the prototype of such food is yogurt, the food market is filled with probiotic-enriched products. In contrast to lyophilized preparations, however, the claims of the effects on health by probiotic-enriched foods are rarely supported by solid evidence.

Probiotics are given as therapy both as an adjunct to other treatments and as primary therapy. Conclusive proof of efficacy by probiotic therapy has been reached in acute childhood gastroenteritis. In other conditions, such as obesity, the role of probiotics is less clear or limited to specific settings. However, evidence of efficacy is accumulating in several conditions, affecting either the intestine or nonintestinal organs.

Finally, the concept that the benefits of probiotics are only effective for minor conditions is changing and today neonatologists are challenged with recommendations to use probiotics in preterm babies with the aim of reducing the incidence of necrotizing enterocolitis as well as death, independent of necrotizing enterocolitis in very-low-birth-weight newborns. This indication is paralleled by data indicating the role of the mother's microflora during pregnancy in the immune programming of the child and the risk of atopy.

Overall, probiotics appear capable of affecting a number of functions and conditions, which is not without major commercial consequences. The risk is – on one hand – to regard probiotics as 'generally good and able to produce mild beneficial effects to virtually everyone with any disease'. Actually, however, the opposite is true, i.e. 'selected probiotic preparations are effective in selected conditions among specific populations'. Further high-level research along this path and rigorous information to interested parties (physicians, patients, customers) will ultimately result in a major benefit for everybody.

This is exactly the purpose of this book: to provide unbiased, updated information on several exciting developments in biology, pharmacology, and medicine in this rapidly evolving scenario which is progressively having major consequences on our knowledge and actions.

Alfredo Guarino, Naples
Eamonn M.M. Quigley, Cork
W. Allan Walker, Boston

Guarino A, Quigley EMM, Walker WA (eds): Probiotic Bacteria and Their Effect on Human Health and Well-Being.
World Rev Nutr Diet. Basel, Karger, 2013, vol 107, pp 1–8 (DOI: 10.1159/000345729)

Probiotics: Definition and Taxonomy
10 Years after the FAO/WHO Guidelines

Lorenzo Morelli

Facoltà di Agraria, UCSC Piacenza e Cremona, Piacenza, and Istituto di Microbiologia e Centro Ricerche
Biotecnologiche, Cremona, Italy

Abstract

Between 2001 and 2002, two joint FAO and WHO documents were published with the aim of providing guidelines for assessing the safety and efficacy of the bacteria generally defined as 'probiotics' that
are used for providing beneficial health effects in humans. These documents were reports of two international meetings of experts convened by the FAO and WHO as a response to the booming activity
in both research and product innovation. Ten years later, it is now possible to examine the impact of
these documents on the international regulatory framework as well as on the academic world. The
most relevant impact is related to the definition of probiotics, which is now used all over the world as
well as gained consideration in national legislations. However, there is still a misuse of the definition
due to confusion between the use of beneficial bacteria in food and pharmaceutical-like products. In
the European Union, the FAO/WHO documents are currently used as a regulation benchmark for the
evaluations by the European Food Safety Authority. In the meantime, the need for a correct and inter-
nationally acknowledged taxonomic identification, as required by these documents, has been accepted by both scientists and regulatory authorities, improving the quality of the area. A positive evaluation
of the impact of these guidelines can then be done after 10 years, even if their use is still partial.

A quick analysis of data retrievable from the major bibliographic databases such as
PubMed, Scopus, and ISI – Web of Science clearly shows that the use of the term
'probiotic(s)' as a keyword has shown a constant increase over the last 20 years. In the
PubMed database, it appeared about a dozen times per year in the 1990s, increasing
to hundreds of times per year in the 2000s, and is now showing up in the thousands
range (www.gopubmed.org). This blend of Latin (pro = for, in favor of) and Greek
(bios = life) was originally intended to define substances able to support the growth
of microorganisms [1], but later refined to identify 'a live microbial feed supplement
which beneficially affects the host animal by improving its intestinal balance' [2].

In the last three decades of the previous century there was, at least in Europe, a great scientific interest in trying to replace the so-called growth promoter or auxinic antibiotics in animal feed with ingredients that have no potential to raise resistance to antibiotics used in veterinary medicine and/or humans. Further modifications were introduced later on, reshaping the whole meaning of the definition. Guarner and Schaafsma [3] defined probiotics as: 'live microorganisms, which when consumed in adequate amounts, confer a health effect on the host'. These authors had in mind the use of beneficial bacteria in humans and introduced two novel aspects in their definition: (1) the need to provide a certain amount of viable bacterial cells to obtain a health effect, and (2) the health effect is not necessarily related to an action of the gut microbiota.

At the end of 1990s a considerable amount of new data on the mechanisms responsible for the probiotic mechanisms, such as the modulation of the gut-associated action immune system, became available in the scientific literature. In addition, the European Union began to financially support projects devoted to characterizing functional foods. Private companies were also funding research devoted to patenting and industrially exploiting beneficial bacteria as active ingredients for fermented milks or dairy probiotic products. These foods aimed to provide a specific health benefit to consumers not related to their nutritional content. From a regulation point of view, however, they were a gray area with a lack of international consensus on their safety as well as on the methodology used to measure their efficiency in delivering to consumers the claimed beneficial effects.

At the beginning of the 2000s, the FAO and WHO undertook an evaluation process in order to provide international guidelines on functional and safety aspects of probiotics. 2001 saw the first results of their efforts when an expert group was jointly convened by the FAO/WHO in Córdoba (Argentina) [4] with the aim of discussing the health and nutritional properties of powder milk with live lactic acid bacteria. It is noteworthy to point out that the government of Argentina requested the meeting in order to solve a commercial dispute, and the meeting served to provide additional evidence of the need to support the results of the innovation process which had taken place in the food area within a new regulatory framework.

Eleven experts (including myself) from 10 countries participated in the meeting organized by Dr. M. Pineiro (FAO) and J. Schlundt (WHO). The group had the mission of reviewing and assessing the available science on the supplementation of milk for human consumption with viable lactic acid bacteria, with a special emphasis on powdered milk supplemented with live lactic acid bacteria. The experts were then asked to discuss the tools used in safety and nutritional assessment of probiotics as well as to review the scientific basis for the efficacy evaluation and potential health claims. A final request to the experts was aimed at obtaining recommendations on future needs in this area, including research priorities for assessing the safety and efficacy of probiotics.

A second step of this joint FAO/WHO joint effort was done in 2002 [5] when the results of a meeting of experts held in London (Ont., Canada) were summarized in 'Guidelines for the evaluation of probiotics in food', in which some details about the tools to be used for identification, characterization, and efficacy demonstration were provided. Now, 10 years later, it is time to evaluate the impact of these documents, which is the purpose of this chapter.

The 2001 Document: Definition(s)

The first document [4] is always cited for the definition of probiotics: 'Live microorganisms which when consumed in adequate amounts as part of food confer a health benefit on the host.' However, little or no attention is generally paid to the lines immediately before and after this definition. It should be remembered that there are two definitions provided by this document, not just one. In the very next lines, the expert group restricts its scope and the definition with a footnote additionally stating that 'Water is included as a food.' It also needs to be pointed out that the group clearly reports to 'exclude reference to the term biotherapeutic agents, and beneficial microorganisms not used in food.'

It can thus be concluded that while the term 'probiotic' can be used to cover the use of beneficial bacteria in general and not only in food, the guidelines provided by the expert group deal with their use in food only. This subtle but extremely relevant difference implies that the 2001 document was directed at the administration of beneficial bacteria to 'otherwise healthy people' and was not covering their use as pharmaceutical products. In fact, section 5.3.6. of the document is titled 'Use of probiotics in otherwise healthy people' and states: 'Many probiotic products are used by consumers who regard themselves as being otherwise healthy. They do so on the assumption that probiotics can retain their health and well-being, and potentially reduce their long-term risk of diseases of the bowel, kidney, respiratory tract and heart. Several points need to be made on this assumption and its implications.' The section concludes by stating: 'Consultation would like studies to be done to give credibility to the perception that probiotics should be taken on a regular basis by healthy men, women and children.' Such a conclusion was ignored by both scientists and industry, which in the following years produced a large amount of data, mainly positive, obtained in clinical trials performed on patients and not with healthy people.

Beneficial bacteria can be utilized to treat pathological conditions, but in these cases the pharmaceutical rules apply, as the word probiotic is clearly used by the FAO/WHO experts to identify beneficial bacteria not used with the intention to treat, whatever the pathology, even if the 2001 document provides a long list of positive results obtained by *treating* pathological conditions with bacteria.

This difference is of paramount relevance for problems such as safety assessment. For food ingredients, a 'long history of safe use' is enough, but for drugs it is mandatory to assess safety through a complex system including in vitro tests, animal models, and ultimately human volunteers. This is the reason why the sections devoted to *safe-*

ty of probiotics are so short in both FAO/WHO documents. The two key sentences of section 7.2 'Safety in humans' state: 'Information acquired to date shows that lactobacilli have a long history of use as probiotics without established risk to humans, and this remains the best proof of their safety. Also, no pathogenic or virulence properties have been found for lactobacilli, bifidobacteria or lactococci.'

The expert group then looked at the need to assess the presence of antibiotic resistance, an evaluation done at the strain level, while the overall safety was then linked to the species and its long history of safe use in foods for *healthy* humans. The misuse has caused and is still causing problems: bacteria considered to be safe on the basis of their historical presence in foods have been used to *treat* patients affected by severe pathologies, sometimes causing threats to their health [5, 6] and raising clinicians' doubts about the safety of probiotics and the FAO/WHO approach to safety assessment [6].

However, the use of bacteria in pathological conditions is not included in the scope of the FAO/WHO documents; therefore, in these documents the safety of bacteria is granted on the basis of a 'long history of safe use in food' and thus a use in healthy people but not in patients. For the latter, safety assessment has to follow the procedures which are mandatory for drugs. The use of probiotics in patients without a proper safety assessment is then due to the misunderstanding or ignorance of clinicians concerning all of the qualifications of the definitions provided by the FAO and WHO. As a final remark, although the 2001 document has reached a wide diffusion, 10 years later its knowledge seems limited to only some selected sentences.

The 2002 Guidelines: Relevance of a Correct Taxonomy

The second document [7], which is much shorter than the first, focuses on the technical aspects of the correct taxonomic identification, as well as on the procedures to be used for the characterization of probiotics and the demonstration of their efficacy. This document is also characterized by a scheme summarizing the steps of the process needed for qualifying a bacterium as 'probiotic' (table 1).

The process starts with the correct identification of the probiotic bacterium at the strain level. This is of paramount relevance since the beneficial properties are strain-specific. It also enables accurate postmarketing surveillance and epidemiological studies for safety/efficacy purposes.

It is clear that a correct taxonomy is required in order to ensure that allocation of the probiotic strain to a species with a tradition of safe use in food. Concerning efficacy, the 2002 document suggested only one possible exception to the strain-related effect: the well-recognized ability of all strains of *Streptococcus thermophilus* and *Lactobacillus delbrueckii* ssp. *bulgaricus* to enhance lactose digestion in lactose-intolerant individuals, a beneficial effect not related to the strain but instead to the species. This

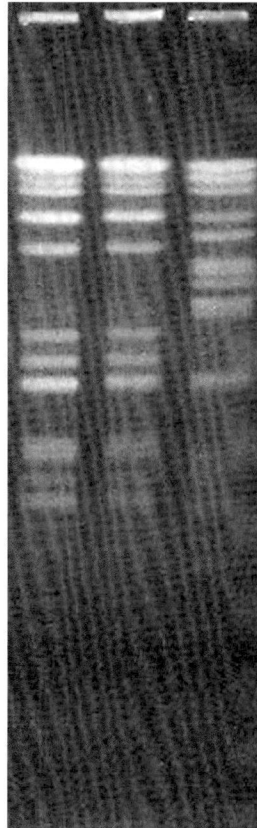

Fig. 1. Pulse field gel electrophoresis of 3 strains of *Lactobacillus paracasei:* in the left and middle lanes two strains with an identical pattern, at right a different strain.

Table 1. Steps of the process to be followed in order to qualify a bacterium as probiotic according to the FAO/WHO guidelines (2002)

Identification	Strain identification by phenotypic and genotypic methods at the genus and species level
	Deposit strain in international culture collection
Safety assessment	In vitro tests and/or animal models
	Phase 1 human study
	Antibiotic resistance pattern
Efficacy assessment	In vitro tests
	Animal studies
	Double-blind, randomized, placebo-controlled (DBPC) phase 2 human trial or other appropriate design with sample size and primary outcome appropriate for determining whether the strain/product is efficacious
	Preferably a second independent DBPC study to confirm results

Table 2. Identification schemes proposed by the FAO/WHO in 2001 and 2002 and by the EFSA in 2008

	FAO/WHO	EFSA
Species identification	DNA-DNA hybridization DNA sequences encoding 16S rRNA	DNA sequences encoding 16S rRNA
Strain identification	Pulse field gel electrophoresis (see fig. 1) Randomly amplified polymorphic DNA	Pulse field gel electrophoresis Randomly amplified polymorphic DNA
Also suggested	Sugar fermentation pattern, plasmid profiling, deposit in an internationally recognized culture collection	

exception was positively considered by the European Food Safety Authority (EFSA) when the NDA panel issued its opinion [8] in favor of the use of health claims on the efficacy of yogurt cultures to reduce lactose intolerance.

An additional application of the FAO/WHO guidelines was the endorsement by the EFSA [9] of the identification scheme (table 2) proposed by this document.

It is also surprising that even though these guidelines were released in 2002, their impact on the industrial world was very limited since a large number of the applications submitted 5 years later to the EFSA by food companies in order to be allowed to make a health claim in the European Union market were rejected due to an insufficient taxonomic characterization.

Finally, in one of its 'opinions' [9], the EFSA specifically asked for an identification process identical to the one proposed in 2002 by the FAO/WHO.

The Impact: 10 Years Later

The FAO/WHO documents had an impact on both regulations and science in the field. Several countries adopted the definition suggested by the expert group in their regulatory documents:

- Italy – Guidelines of the Italian Ministry of Health (Ministero della Salute), 2005, revised in 2011
- France – dossier from the French Agency for Food Safety (Agence Francaise de Sécuritè Sanitaire des Aliments), 2005
- Canada – Probiotics in Food, Department for Public Health of Canada/Health Canada, 2009
- Argentina – modification to art. 1389 of the 'Codigo Alimentario Argentino', December 12, 2011
- India – Guidelines for Evaluation of Probiotics in Food – Indian Council of Medical Research, Department of Health Research of the Ministry of Health and Family Welfare and Department of Biotechnology, Ministry of Science and Technology, New Delhi, 2011

It is clear that after 10 years the FAO/WHO definition of probiotics is a reference term for the food and food supplement sector as well as a reference for legislation.

In addition, CODEX has also been involved [10] in setting the standard for probiotics, taking advantage of the definition provided by the expert group 10 years ago. This is not true for the academic world. Even if always cited, the word 'probiotic' and its definition by FAO/WHO is often used in the frame of clinical trials and pathological conditions without a clear knowledge of the 'food area' in which the documents were allocated.

In conclusion, attention must be paid to the different concepts underlying the use of beneficial bacteria in food and pharmaceuticals. Bacteria intentionally added to food in order to provide a health benefit ('probiotics' according to the FAO/WHO documents) have a safe status simply based on the long history of their use in 'healthy people'. The same bacteria (even the same bacterial strains) used in clinical settings or addressing pathological conditions (the 'biotherapeutics' according to the same FAO/WHO documents) have to be assessed for their safety in these 'new conditions of use'.

The EFSA's negative assessment of hundreds of applications have pointed out the need to carefully follow the indications provided by regulatory bodies; this is where research must be finalized to provide outcomes useful for supporting *applicable innovations* in the industrial world.

The *probiotic crisis* caused by the EFSA could induce a positive rethinking of the research planned in this area, leading to a clear differentiation between food and pharmaceutical use of bacteria. An impressive amount of data have recently been delivered, using genomic approaches, on the relevance of the microbiota and its microbiome for the human health – the idea of a positive manipulation of the functions of the microbiota is more alive and attractive than before.

Even if the history of probiotics dates back to the beginning of the 900s, it seems more than possible that their full potential to be beneficial to humans has yet to be fully described.

References

1 Lilly DM, Stillwell RH: Probiotics: growth promoting factors produced by microorganisms. Science 1965;147:747–748.
2 Fuller R: Probiotics in man and animals. J Appl Bact 1989;66:365–378.
3 Guarner F, Schaafsma GJ: Probiotics. Int J Food Microbiol 1998;39:237–238.
4 FAO/WHO: Report on Joint FAO/WHO Expert Consultation on Evaluation of Health and Nutritional Properties of Probiotics in Food Including Powder Milk with Live Lactic Acid Bacteria. 2001. ftp://ftp.fao.org/es/esn/food/probio_report_en.pdf.3.
5 Besselink MG, van Santvoort HC, Buskens E, Boermeester MA, van Goor H, Timmerman HM, Nieuwenhuijs VB, Bollen TL, van Ramshorst B, Witteman BJ, Rosman C, Ploeg RJ, Brink MA, Schaapherder AF, Dejong CH, Wahab PJ, van Laarhoven CJ, van der Harst E, van Eijck CH, Cuesta MA, Akkermans LM, Gooszen HG, Dutch Acute Pancreatitis Study Group: Probiotic prophylaxis in predicted severe acute pancreatitis: a randomised, double-blind, placebo-controlled trial. Lancet 2008;371:651–659.

6 Kochan P, Chmielarczyk A, Szymaniak L, Brykczyn-ski M, Galant K, Zych A, Pakosz K, Giedrys-Kalemba S, Lenouvel E, Heczko PB: *Lactobacillus rhamnosus* administration causes sepsis in a cardiosurgical patient – is the time right to revise probiotic safety guidelines? Clin Microbiol Infect 2011;10:1589–1592.

7 FAO/WHO: Report of a Joint FAO/WHO Working Group on Drafting Guidelines for the Evaluation of Probiotics in Food. 2002. ftp://ftp.fao.org/es/esn/food/wgreport2.pdf.

8 European Food Safety Authority. EFSA J 2010;8:1763.

9 European Food Safety Authority. EFSA J 2009;7:1247.

10 Pineiro M, Stanton C: Probiotic bacteria: legislative framework – requirements to evidence basis. J Nutr 2007;137(3 suppl 2):850S–853S.

Lorenzo Morelli
Preside/Dean Facoltà di Agraria, UCSC Piacenza e Cremona
Via Emilia Parmense 84
IT–29122 Piacenza (Italy)
E-Mail lorenzo.morelli@unicatt.it

Guarino A, Quigley EMM, Walker WA (eds): Probiotic Bacteria and Their Effect on Human Health and Well-Being.
World Rev Nutr Diet. Basel, Karger, 2013, vol 107, pp 9–16 (DOI: 10.1159/000345732)

Intestinal Microbiota Composition in Children

Vittoria Buccigrossi · Emanuele Nicastro · Alfredo Guarino

Department of Translational Medical Science, Section of Pediatrics, University of Naples Federico II,
Naples, Italy

Abstract

Culture-independent strategies such as high-throughput parallel sequencing and comparative genomics, metabolic profiling and functional genomics, fluorescence in situ hybridization, and phylogenetic microarray are providing new insights into the composition, determinants, and functional
roles of human microbiota. The initial colonization and subsequent development of the intestinal
microbiota in early life is a key step for the composition of the human microbiota throughout one's
life. An increasing body of evidence highlights the role of nutrition both in early infancy and childhood in the age-related development of the microflora, but other factors significantly contribute to
the final composition in the adult age, such as antibiotics and/or functional foods, both of which are
widely used in children. The microflora in children is plastic and susceptible to changes in response
to diet modifications, antibiotic treatment, and other events. In contrast, the microbiota in adults is
relatively stable, reflecting the resilience of the microbiome to external perturbations, such as antibiotic therapies. Eubiosis, i.e. a normal microflora structure, provides protection against infections,
educates the immune system, ensures tolerance to foods, and contributes to nutrient digestion and
energy harvest. However, changes in the microflora, i.e. the presence of too many harmful species
or not enough commensal species, called dysbiosis, produce dysfunctions such as intestinal inflammation. It is becoming clear that dysbiosis plays a role in a broad range of important intestinal and
extraintestinal diseases such as inflammatory bowel diseases, atopy, and obesity. Moreover, abnormal patterns of microflora have been consistently detected in specific diseases and may provide
biomarkers to monitor their course.   Copyright © 2013 S. Karger AG, Basel

The human intestinal tract harbors a complex microbial ecosystem consisting of
about 10–100 trillion bacteria defined as 'gut microbiota', representing 10 times
more bacterial cells than the number of cells in the human body. The majority of
gut bacterial phylotypes belongs to a limited set of phyla such as Firmicutes, Bacteroidetes, Actinobacteria, Proteobacteria, and Verrucomicrobia; however, recent estimates suggest that 800–1,000 different bacterial species inhabit the gastrointestinal

tract, and the set of microbial genomes represents more than 100 times the human genome [1].

The knowledge of the gut microbiota composition, how it interacts with the host, and how it causes human disease has been enhanced by advances in culture-independent techniques for its phylogenetic investigation and quantification. The biggest challenge that researchers have set for themselves is to understand the functional relationship between microorganisms and the host in order to identify their role in diseases. The research in the gut microbiome has undergone an exponential increase in the last 20 years as can be seen when the key words 'intestinal microflora' are searched for in the Web of Knowledge citation database (http://apps.webofknowledge.com).

Techniques to Study the Intestinal Microflora

Molecular methods allow cultivation-independent characterization of gut microbial communities, increasing the ability to identify single species microorganisms and their amounts in fecal samples [2].

Most molecular studies have investigated the sequence of the small subunit ribosomal RNA (16S rRNA) gene [3]. In order to study the diversity of intestinal microflora, the three approaches most commonly used are gene-targeting polymerase chain reaction (PCR) techniques, oligonucleotide probe-based hybridization techniques such as FISH (fluorescent in situ hybridization), and molecular fingerprinting techniques such as DGGE (denaturing gradient gel electrophoresis) [3].

Gene-targeting techniques use gene-specific primers to specifically amplify target genes, including conserved 16S rRNA genes. This approach has been widely applied to gut microbiota analysis and has revealed substantial bacterial diversity and groups of unculturable microbes [4]. In addition, reverse transcriptase PCR from RNA has also been used to clone genes from fecal samples.

One of the methods to analyze intestinal microbiota is rRNA FISH. Based on the rRNA sequences, probes can be synthesized to bind specifically to organisms of interest. Using probes labeled with different fluorescent dyes, different types of microbes can simultaneously be visualized within complex communities [5]. The advantages of FISH are its simple methodology, reduced working time, and low cost. Recently, a new application of FISH was described that extends our understanding of the pathogen-intestine interaction to the spatial localization of bacteria in stools [5].

For microbial diversity analysis, molecular fingerprinting techniques are usually used to analyze the sequence of 16S rRNA from different microbial species. DGGE and TGGE (temperature gradient gel electrophoresis) are perhaps the most commonly used techniques and have been helpful in providing an overview of microbial diversity in the gut symbiotic microbiota. Roudière et al. [6] recently optimized these techniques to assess the diversity of gut microbiota in neonates. Even though these molecular fingerprinting techniques have revealed significant microbial diversity in fecal samples, they provide rather limited information on microbial diversity and the recently developed metagenomic platforms are rapidly replacing these molecular fingerprinting techniques.

Metagenomics explores the global composition of microbial communities, combining the molecular biology and genetic techniques to identify and characterize genetic material from complex microbial environments [7]. DNA microarray technology is a powerful tool for rapid high throughput detection of thousands of 16S ribosomal gene sequences, providing a profile of microbial communities in the human gut. A microarray consisting of 40-mer probes has been developed, tested, and optimized to detect a range of bacterial taxonomic hierarchies. The microarray approach was used to study the microflora in the adult human gastrointestinal tract and to monitor the development of the human infant intestinal microbiota [8]. However, this technique identifies the diversity, but not the relative numbers of each species in the fecal sample, and its major limit is the lack of established clinical correlates, which may create problems in the interpretation of results.

From Birth to the Adult Age: Composition and Dynamics of Intestinal Microflora

The human host-microbe symbiosis starts at birth; however, a study performed by DiGiulio et al. [9] demonstrated the presence of bacteria in amniotic fluid, suggesting that the newborn gut may not be sterile at birth.

Bacteria from the mother's intestinal and vaginal sites, and from the outer environment, colonize the neonatal gut within a few hours after birth. The intestinal microflora of infants delivered via cesarean section is less diverse in terms of bacteria species, with lower bacterial numbers than the microbiota of infants born vaginally, and it is associated with a predominance of staphylococci, corynebacteria, propionibacteria, and Clostridia compared with a microflora dominated by lactobacilli, *Prevotella*, *Bacteroides* and bifidobacteria in vaginally born infants [1].

The time of bacterial colonization is also important and preterm neonates show a delayed bacterial colonization and a limited number of colonizing species [10]. An increasing diversity of gut microflora was observed in the first 8 weeks of life in preterm infants and most infants had staphylococci in their stools as the main species, whereas few infants were colonized with *Bifidobacterium* spp. [10]. This study also showed a positive relationship between the diversity of intestinal microflora and nutrient tolerance and weight gain, supporting a functional relationship between humans and their intestinal microbes.

In the first 48 h after birth, infants are colonized by facultative anaerobic bacteria such as *Escherichia coli* and *Streptococci*. Thereafter, the colonization by *Enterococcus* and *Lactobacillus* species takes place, generating an anaerobic environment by consuming oxygen. At 1 week of age, *Bifidobacterium*, *Bacteroides*, and *Clostridium* are detected in the feces. Also at this time, *Bifidobacterium* species become dominant in human milk-fed infants, usually in parallel with a decrease in Enterobacteriaceae. The presence of other bacterial species, such as *Ruminococcus*, is also important in the early colonization process, as ruminococci are able to inhibit the colonization of Clostridia by the production of ruminococcin A [11].

Nutrition plays a major role at birth and after birth. Breast-fed infants had twice the number of bifidobacteria than formula-fed infants, and in the latter *Atopobium* and *Bacteroides* were found in significant counts. Moreover, the intestinal microbiota was less complex (or 'diverse') in formula-fed infants than in breast-fed infants [12]. This difference is due to the presence of nondigestible oligosaccharide in human milk, which is known as a stimulating factor for bifidobacteria growth [13]. This prebiotic effect can be considered as a major health benefit because it shapes the microbiota, which is believed to stimulate the developing immune and metabolic system.

Breast milk is important since it not only provides a range of substrates for bacterial growth, but is also a natural bacterial inoculum which affects neonatal colonization including staphylococci, streptococci, bifidobacteria, and lactic acid bacteria [14]. Later in life, weaning is associated with the transition toward an adult-like microbiota and the introduction of foods such as cereals, fruits, and vegetables provide a new source of nondigestible carbohydrates [15]. The counts of *Bacteroides*, *Clostridium*, and anaerobic streptococci increase after the introduction of solid weaning foods, whereas facultative anaerobes, such as bifidobacteria and lactobacilli, remain high [16]. Overall, the dynamics of the intestinal microflora follows an unclear timeline since several factors, such as the use of antibiotics, functional nutrients and hygiene status, influence the final bacterial composition.

Although it is generally thought that the gut microbiome in adolescents is similar to that of adults, a recent study showed that bacterial gut communities are substantially different between groups of 11- to 18-year-olds and 22- to 61-year-olds. The main gaps in relative abundances were found for Clostridia, Actinobacteria, and β- and γ-Proteobacteria [17].

The adult microbiota is relatively stable [18]. The temporal stability reflects the resilience of the microbiome in adults, and perturbations, such as antibiotic therapies, have only transient effects on the dominant microbiota. Overall, the adult microbiome is more complex than its infant equivalent, and is stable over time and similar between individuals [19]. In contrast, the gut microbiota in infancy possesses a relatively simple structure, but is rather unstable over time. The microflora in children is intrinsically plastic and affected by few variables.

Recently, Arumugam et al. [20] described the enterotypes as a group of microbial groups that together contribute to build a preferred ecosystem or type of gut microbiota. The characterization of the enterotypes has revealed that an adult human being can be divided into three categories according to their microbiota. The enterotype defines the quantitatively dominating taxa. Enterotype 1 is dominated by *Bacteroides*, enterotype 2 by *Prevotella*, and enterotype 3 by *Ruminococcus,* and each of the three enterotypes differently affects the host metabolic functions [20]. The enterotypes are associated with protein and animal fat *(Bacteroides)* or a carbohydrate-rich diet *(Prevotella)*, and may change as early as within 24 h of initiating a high-fat/low-fiber or low-fat/high-fiber diet [21]. However, the core structure of enterotypes does not change and is a lifestyle hallmark of individuals.

The relationship between diet and the intestinal microflora structure emerged from a comparative evaluation of bacteria in European children (EU), who ate a typical Western diet high in animal protein and fat, and children in Burkina Faso, who were on high carbohydrate/low animal protein diet [22]. The African children showed a significant enrichment in Bacteroidetes and depletion in Firmicutes, with an abundance of *Prevotella* and *Xylanibacter,* known to hydrolyze cellulose and xylan which are completely lacking in the European children. In addition, more short-chain fatty acids were found in the children from Burkina Faso than in the EU children. The hypothesis is that the gut microbiota is adapted to the polysaccharide-rich diet which allows one to maximize energy intake from fibers and that short-chain fatty acid-producing bacteria help to prevent the establishment of intestinal microbes causing diarrhea, as seen by the fact that Enterobacteriaceae, such as *Shigella* and *Escherichia*, were significantly underrepresented in children from Burkina Faso than in EU children.

Eubiosis and Dysbiosis

A healthy state of microbiota structure, in which microorganisms with potential health benefits predominate in number over those which are potentially harmful, is defined as 'normobiosis' (or 'eubiosis'). On the contrary, 'dysbiosis' is a condition in which one or more potentially harmful bacterial species are dominant [23].

A healthy microflora is characterized by its richness and evenness. 'Richness' reflects the number of bacterial species in a specific ecosystem, not taking into account their relative abundance. 'Evenness' indicates the relative abundance of each species in a specific ecosystem. These two definitions are used to describe the microbial diversity in the gastrointestinal tract [24].

In many diseases, the diversity of the microflora is reduced. However, there are also specific aberrations of the microflora in selected childhood diseases. Consistent abnormalities in the microbial structure have been detected in populations of children with specific diseases and have been defined as 'microbiological signatures' of those diseases. The word 'signature' provides a double concept: (1) the specific role of selected bacterial species in causing – or contributing to – specific diseases, and (2) the opportunity to recognize the disease (or monitoring its course) by analyzing microflora composition, which means that microbial aberrations may be a hallmark of that disease. However, the changes observed in microbial populations raise the option of targeting microflora for therapy. This is being done with increasing success in selected diseases using various strategies, including administration of probiotics. Several examples are provided by chronic diseases such as celiac disease in which children show a peculiar microbial pattern with abundant Firmicutes [25] or in inflammatory bowel diseases where the presence of *Faecalibacterium prausnitzii* is inversely correlated with the severity of disease in acute appendicitis [26]. Irritable bowel syndrome is associated with dysbiosis, and the manipulation of intestinal microbial communities (i.e. probiotics) may effectively alleviate fastidious symptoms. Interestingly, a close relationship between a specific bacterial profile with the severity of symptoms was recently shown in pediatric irritable bowel syndrome [27].

A major insight into the role of the microflora and the chance to target it for therapy is necrotizing enterocolitis (NEC), a severe and potentially fatal disease which affects preterm neonates. A decrease in microbiota diversity was observed for all preterm infants, and NEC children showed a further reduction in diversity, with a predominance of γ-Proteobacteria and a reduction of other bacterial species [28]. However, other researchers have found an opposite pattern, and in a prospective study, bacterial diversity expressed as band richness was higher in NEC than in controls [29]. In randomized controlled trials the use of probiotics was effective in decreasing the risk of NEC and the global mortality in preterm infants [30]. This observation supports the concept about the identification of a pattern of microbiota at high-risk of NEC.

Conclusion

Compared to the complex and resilient microbiome in adults, intestinal microflora in children is intrinsically plastic and deeply affected by few variables, but less exposed to factors that may change its composition. Generally, a reduction in the diversity of gut microflora is associated with intestinal inflammation [31], and specific aberrations in microflora composition are microbial 'signatures' of selected diseases, suggesting a specific role of well-defined bacterial species in the pathophysiology of those diseases or also their use as biomarkers to predict them or to monitor their course. Finally, a considerable amount of research is focused on modifying the intestinal flora with probiotic bacteria to attenuate intestinal inflammation and prevent infantile gastroenteritis [32].

References

1 Scholtens PA, Oozeer R, Martin R, Amor KB, Knol J: The early settlers: intestinal microbiology in early life. Annu Rev Food Sci Technol 2012;3:425–447.

2 Guarino A, Giannattasio A: New molecular approaches in the diagnosis of acute diarrhea: advantages for clinicians and researchers. Curr Opin Gastroenterol 2011;27:24–29.

3 Su C, Lei L, Duan Y, Zhang K-Q, Yang J: Culture-independent methods for studying environmental microorganisms: methods, application, and perspective. Appl Microbiol Biotechnol 2012;93:993–1003.

4 Karlsson CLJ, Molin G, Cilio CM, Ahrné S: The pioneer gut microbiota in human neonates vaginally born at term – a pilot study. Pediatr Res 2011;70:282–286.

5 Swidsinski A, Loening-Baucke V, Herber A: Mucosal flora in Crohn's disease and ulcerative colitis – an overview. J Physiol Pharmacol 2009;60:61–71.

6 Roudière L, Jacquot A, Marchandin H, Aujoulat F, Devine R, Zorgniotti I, Jean-Pierre H, Picaud J-C, Jumas-Bilak E: Optimized PCR-temporal temperature gel electrophoresis compared to cultivation to assess diversity of gut microbiota in neonates. J Microbiol Methods 2009;79:156–165.

7 Tringe SG, von Mering C, Kobayashi A, Salamov AA, Chen K, Chang HW, Podar M, Short JM, Mathur EJ, Detter JC, Bork P, Hugenholtz P, Rubin EM: Comparative metagenomics of microbial communities. Science 2005;308:554–557.

8 Palmer C, Bik EM, DiGiulio DB, Relman DA, Brown PO: Development of the human infant intestinal microbiota. PLoS Biol 2007;5:e177.

9 DiGiulio DB, Romero R, Amogan HP, Kusanovic JP, Bik EM, Gotsch F, Kim CJ, Erez O, Edwin S, Relman DA: Microbial prevalence, diversity and abundance in amniotic fluid during preterm labor: a molecular and culture-based investigation. PLoS One 2008;3:e3056.

10 Jacquot A, Neveu D, Aujoulat F, Mercier G, Marchandin H, Jumas-Bilak E, Picaud J-C: Dynamics and clinical evolution of bacterial gut microflora in extremely premature patients. J Pediatr 2011;158:390–396.

11 Morelli L: Postnatal development of intestinal microflora as influenced by infant nutrition. J Nutr 2008;138:1791S–1795S.

12 Bezirtzoglou E, Tsiotsias A, Welling GW: Microbiota profile in feces of breast- and formula-fed newborns by using fluorescence in situ hybridization (FISH). Anaerobe 2011;17:478–482.

13 Thurl S, Munzert M, Henker J, Boehm G, Müller-Werner B, Jelinek J, Stahl B: Variation of human milk oligosaccharides in relation to milk groups and lactational periods. Br J Nutr 2010;104:1261–1271.

14 Albesharat R, Ehrmann MA, Korakli M, Yazaji S, Vogel RF: Phenotypic and genotypic analyses of lactic acid bacteria in local fermented food, breast milk and faeces of mothers and their babies. Syst Appl Microbiol 2011;34:148–155.

15 Koenig JE, Spor A, Scalfone N, Fricker AD, Stombaugh J, Knight R, Angenent LT, Ley RE: Succession of microbial consortia in the developing infant gut microbiome. Proc Natl Acad Sci USA 2011;108:4578–4585.

16 Amarri S, Benatti F, Callegari ML, Shahkhalili Y, Chauffard F, Rochat F, Acheson KJ, Hager C, Benyacoub J, Galli E, Rebecchi A: Changes of gut microbiota and immune markers during the complementary feeding period in healthy breast-fed infants. J Pediatr Gastroenterol Nutr 2006;42:488–495.

17 Agans R, Rigsbee L, Kenche H, Michail S, Khamis HJ, Paliy O: Distal gut microbiota of adolescent children is different from that of adults. FEMS Microbiol Ecol 2011;77:404–412.

18 Jalanka-Tuovinen J, Salonen A, Nikkilä J, Immonen O, Kekkonen R, Lahti L, Palva A, de Vos WM: Intestinal microbiota in healthy adults: temporal analysis reveals individual and common core and relation to intestinal symptoms. PloS One 2011;6:e23035.

19 Claesson MJ, Cusack S, O'Sullivan O, Greene-Diniz R, de Weerd H, Flannery E, Marchesi JR, Falush D, Dinan T, Fitzgerald G, et al: Composition, variability, and temporal stability of the intestinal microbiota of the elderly. Proc Natl Acad Sci USA 2011;108:4586–4591.

20 Arumugam M, Raes J, Pelletier E, Le Paslier D, Yamada T, Mende DR, Fernandes GR, Tap J, Bruls T, Batto J-M, et al: Enterotypes of the human gut microbiome. Nature 2011;473:174–180.

21 Wu GD, Chen J, Hoffmann C, Bittinger K, Chen YY, Keilbaugh SA, Bewtra M, Knights D, Walters WA, Knight R, Sinha R, Gilroy E, Gupta K, Baldassano R, Nessel L, Li H, Bushman FD, Lewis JD: Linking long-term dietary patterns with gut microbial enterotypes. Science 2011;334:105–108.

22 De Filippo C, Cavalieri D, Di Paola M, Ramazzotti M, Poullet JB, Massart S, Collini S, Pieraccini G, Lionetti P: Impact of diet in shaping gut microbiota revealed by a comparative study in children from Europe and rural Africa. Proc Natl Acad Sci USA 2010;107:14691–14696.

23 Roberfroid M, Gibson GR, Hoyles L, McCartney AL, Rastall R, Rowland I, Wolvers D, Watzl B, Szajewska H, Stahl B, Guarner F, Respondek F, Whelan K, Coxam V, Davicco M-J, Léotoing L, Wittrant Y, Delzenne NM, Cani PD, Neyrinck AM, Meheust A: Prebiotic effects: metabolic and health benefits. Br J Nutr 2010;104:S1–S63.

24 Gerritsen J, Smidt H, Rijkers GT, de Vos WM: Intestinal microbiota in human health and disease: the impact of probiotics. Genes Nutr 2011;6:209–240.

25 Sellitto M, Bai G, Serena G, Fricke WF, Sturgeon C, Gajer P, White JR, Koenig SSK, Sakamoto J, Boothe D, Gicquelais R, Kryszak D, Puppa E, Catassi C, Ravel J, Fasano A: Proof of concept of microbiome-metabolome analysis and delayed gluten exposure on celiac disease autoimmunity in genetically at-risk infants. PloS One 2012;7:e33387.

26 Schwiertz A, Jacobi M, Frick J-S, Richter M, Rusch K, Köhler H: Microbiota in pediatric inflammatory bowel disease. J Pediatr 2010;157:240–244.

27 Saulnier DM, Riehle K, Mistretta T-A, Diaz M-A, Mandal D, Raza S, Weidler EM, Qin X, Coarfa C, Milosavljevic A, Petrosino JF, Highlander S, Gibbs R, Lynch SV, Shulman RJ, Versalovic J: Gastrointestinal microbiome signatures of pediatric patients with irritable bowel syndrome. Gastroenterology 2011; 141:1782–1791.

28 Wang Y, Hoenig JD, Malin KJ, Qamar S, Petrof EO, Sun J, Antonopoulos DA, Chang EB, Claud EC: 16S rRNA gene-based analysis of fecal microbiota from preterm infants with and without necrotizing enterocolitis. ISME J 2009;3:944–954.

29 Smith B, Bodé S, Skov TH, Mirsepasi H, Greisen G, Krogfelt KA: Investigation of the early intestinal microflora in premature infants with/without necrotizing enterocolitis using two different methods. Pediatr Res 2012;71:115–120.

30 Mihatsch WA, Braegger CP, Decsi T, Kolacek S, Lanzinger H, Mayer B, Moreno LA, Pohlandt F, Puntis J, Shamir R, Stadtmüller U, Szajewska H, Turck D, van Goudoever JB: Critical systematic review of the level of evidence for routine use of probiotics for reduction of mortality and prevention of necrotizing enterocolitis and sepsis in preterm infants. Clin Nutr 2012;31:6–15.

31 Guarino A, Wudy A, Basile F, Ruberto E, Buccigrossi V: Composition and roles of intestinal microbiota in children. J Matern Fetal Neonatal Med 2012;25: 63–66.

32 Guarino A, Lo Vecchio A, Berni Canani R: Probiotics as prevention and treatment for diarrhea. Curr Opin Gastroenterol 2009;25:18–23.

Vittoria Buccigrossi
Department of Pediatrics, University Federico II
Via Sergio Pansini 5
IT–80131 Naples (Italy)
E-Mail buccigro@unina.it

Guarino A, Quigley EMM, Walker WA (eds): Probiotic Bacteria and Their Effect on Human Health and Well-Being.
World Rev Nutr Diet. Basel, Karger, 2013, vol 107, pp 17–24 (DOI: 10.1159/000346875)

Intestinal Microbiota Composition in Adults

Virginia Robles Alonso · Francisco Guarner

Digestive System Research Unit, University Hospital Vall d'Hebron, Ciberehd, Barcelona, Spain

Abstract

New sequencing technologies together with the development of bioinformatics allow a description of the full spectrum of the microbial communities that inhabit the human intestinal tract, as well as their functional contributions to host health. Most community members belong to the domain Bacteria, but Archaea, eukaryotes (yeasts and protists), and viruses are also present. Only 7–9 of the 55 known divisions or phyla of the domain Bacteria are detected in fecal or mucosal samples from the human gut. Most taxa belong to just two divisions, Bacteroidetes and Firmicutes, but other divisions that have been consistently found are Proteobacteria, Actinobacteria, Fusobacteria, and Verrucomicrobia. *Bacteroides, Faecalibacterium,* and *Bifidobacterium* are the most abundant genera, but their relative proportion is highly variable across individuals. Full metagenomic analysis has identified more than 5 million nonredundant microbial genes encoding up to 20,000 biological functions related with life in the intestinal habitat. The overall structure of predominant genera in the human gut can be assigned to three robust clusters, which are known as 'enterotypes'. Each of the three enterotypes is identifiable by the levels of one of three genera: *Bacteroides* (enterotype 1), *Prevotella* (enterotype 2), and *Ruminococcus* (enterotype 3). This suggests that microbiota variations across individuals are stratified, not continuous. The next steps include the identification of changes that may play a role in certain disease states. A better knowledge of the contributions of microbial symbionts to host health will help in the design of interventions to improve symbiosis and combat disease.

Copyright © 2013 S. Karger AG, Basel

New Technologies in the Field of Metagenomics

The advent of high-throughput sequencing technologies has led to a turning point in our understanding of the microbial colonization of the human gut. Prior to these techniques, the culture approach was not able to provide a reliable profile of all the communities present in a particular ecological niche because of the variable nutritional needs of the different community members and the difficulty to grow them in laboratory media.

High-throughput sequencing technologies are culture-independent methods allowing the characterization of microbial communities as a whole through the analysis

Table 1. Glossary of terms

Dysbiosis: an imbalance of the normal gut microbiota composition.
Enterotype: a classification of the human gut microbial communities into three groups or types on the basis of the bacteriological composition of the ecosystem (diversity and abundance of the predominant genera).
Metagenome: the total genetic content of the combined genomes of the constituents of an ecological community.
Metagenomics: the study of all the genetic material recovered directly from environmental samples, bypassing the need to isolate and culture individual community members.
Microbiome: the collective genome of the microbial symbionts in a host animal.
Microbiota: the collection of microbial communities colonizing a particular ecological niche.
Phylotype: a microbial group defined by 16S rRNA sequence similarity rather than by phenotypic characteristics. A similarity of 97% indicates approximately a species level.
Symbionts: the microbial partners in symbiosis.
Symbiosis: close and persistent interactions between living organisms of different species. Biological interactions may be 'mutualistic', i.e. both partners derive a benefit, 'commensal', i.e. one partner benefits without affecting the others, and 'parasitic', i.e. one benefits while the other is harmed.

of the genetic material present in an environment. Such investigations enable a deep and global description of all the community members and their relative abundance [1]. The new approach has led to the coining of the term 'metagenomics', which is defined as the study of the genetic material recovered directly from environmental samples, thus bypassing the need to isolate and culture individual community members [2]. The metagenome is the collective genetic content of the combined genomes of the constituents of an ecological community (table 1). The most common approach consists on the extraction of DNA from a biological sample, followed by the amplification and sequencing of 16S ribosomal RNA (rRNA) genes in the sample. The 16S rRNA gene is present in all bacteria and contains both conserved and variable regions. Thus, similarities and differences in the sequence of nucleotides of the 16S rRNA gene allow taxonomic identification ranging from the domain and phylum level to the species or strain level. Taxonomic identification is based on comparison of 16S rRNA sequences in the sample with reference sequences in the database. In this way, studies on the 16S rRNA gene provide information about microbial composition and diversity of species in a given sample.

The most powerful molecular approach is not limited to 16S rRNA sequencing, but it addresses all the genetic material in the sample. The decreasing cost and increasing speed of DNA sequencing, coupled with advances in computational analyses of large datasets, have made it feasible to analyze complex mixtures of entire genomes with reasonable coverage. The resulting information describes the collective genetic con-

Fig. 1. Phylogenetic classification and abundance (logarithmic scale) of microbial genes identified in fecal samples from European individuals. The vast majority of gene sequences belong to the domain Bacteria or cannot be classified (unknown). Only low percentages of the reads were classified as Archaea, eukaryotes, or viruses. Data extracted from supplementary files in Arumugam et al. [5].

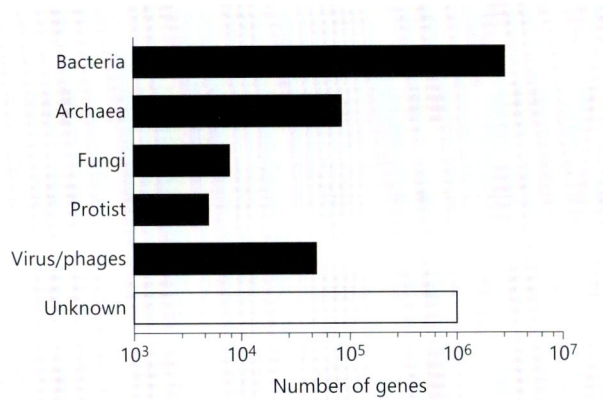

tent of the community from which functional and metabolic networks can be inferred. Thus, the full metagenomic approach has the advantage of not only providing the phylogenetic characterization of the microbial community, but also telling about biological functions present in the community.

In the past few years, two large-scale initiatives of major funding agencies have accomplished the task of deciphering both the structure and function of the human gut microbiota, namely the MetaHIT project – funded by the European Commission – and the Human Microbiome Project – funded by the National Institutes of Health of the USA. Data from these projects have been reviewed and are summarized for the purpose of this chapter.

Diversity and Functions of the Gut Microbiota

Estimates suggest that the colon, by far the largest ecological niche for microbial communities in the human body, harbors over 10^{14} microbial cells, most of them belonging to the domain Bacteria. Methanogenic Archaea, eukaryotes (yeasts and protists), and viruses (phages and animal viruses) are also present (fig. 1). Molecular studies based either on 16S rRNA gene or full metagenomic sequencing have highlighted that only 7–9 of the 55 known divisions or phyla of the domain Bacteria are detected in fecal or mucosal samples from the human gut [3–7]. Moreover, such studies have also revealed that more than 90% of all the bacterial taxa belong to just two divisions: Bacteroidetes and Firmicutes. The other divisions that have been consistently found in samples from the human distal gut are Proteobacteria, Actinobacteria, Fusobacteria, and Verrucomicrobia. Of the 13 divisions of the domain Archaea, only very few species (mostly *Methanobrevibacter smithii*) seem to be represented in the human distal gut microbiota. At a species or strain level, there is considerable variation in the composition of the fecal microbiota among human individuals. Strain diversity between individuals is highly remarkable and studies have found that a large proportion of the identified strain-level phylotypes are unique to each person [3]. Each individual harbors his or her own distinctive pattern of bacterial composition.

Sequencing analysis of the 16S rRNA gene has allowed the description of not only differences in microbial communities between different human subjects, but also intraindividual variability. There are differences between fecal and mucosa-associated communities within the same individual [3]. Bacterial composition in the lumen varies from the cecum to the rectum, and fecal samples do not reflect luminal contents in proximal segments of the gastrointestinal tract. However, the community of mucosa-associated bacteria is highly stable from the terminal ileum to the large bowel in a given individual. Factors such as diet, drug intake, travelling, or simply colonic transit time have an impact on microbial composition in fecal samples over time in a unique host. A recent study [8] collected samples from three different body sites (gut, mouth, skin) of two healthy subjects on a daily basis for a period of 15 and 6 months, respectively. Community differentiation by body site is highly stable over time but, within the same body site, a low stability across time was noticed. At a species level, very few microbial members would constitute the so-called 'core human gut microbiota' since only the most abundant species (around 5% of all species found in fecal samples) were always present in all sampling points from the same individual [8]. Thus, intraindividual variability was shown to be remarkable, but distinction between different individuals was patent and persistent over time.

Deep sequencing analysis of the 16S rRNA gene in fecal samples from a cohort of healthy children and adults from the Amazon region of Venezuela, rural Malawi, and United States metropolitan areas found striking differences in composition and diversity between Westernized and non-Westernized populations [9]. Using principal coordinate analysis, the US samples clustered separately from non-US samples (Malawians and Amerindians), but there was not any significant clustering by village for Malawians and Amerindians or by region within the US. Bacterial diversity increased with age in all three populations and the fecal microbiota of US adults was the least diverse compared with the two other populations, differences being evident in adults and children older than 3 years of age. The differences between US and Malawian/Amerindian microbiotas may be related to several factors including variation in environmental exposure (vertical and horizontal transmission) and dietary habits [9].

Full metagenomic analysis of fecal samples from a cohort of European adult subjects identified a total of 3.3 million nonredundant microbial genes [4]. This effort provided for the first time a gene catalogue of the human gut microbiome. Each individual carried an average of 600,000 nonredundant microbial genes in the gastrointestinal tract (table 2). This figure suggests that most of the 3.3 million genes in the catalogue are shared. It was found that around 300,000 microbial genes were common

Table 2. The human gut metagenome, our other genome

Microbial genes in the human gut	Number of genes
Median gene set per individual	590,384
Common genes (present in at least 50% of individuals)	294,110
Rare genes (present in less than 20% of individuals)	2,375,655

Data from Qin et al. [4].

in the sense that they were present in at least 50% of individuals. Up to 98% of identified genes in the catalogue were bacterial (fig. 1) and the entire cohort of individuals harbored between 1,000 and 1,150 prevalent bacterial species, with at least 160 species per individual [6]. Interestingly, *Bacteroides, Faecalibacterium,* and *Bifidobacterium* were the most abundant genera, but their relative proportion was highly variable across individuals.

The Human Microbiome Project analyzed samples from the digestive tract, mouth, skin, nose, and female urogenital tract of healthy adult individuals [6, 7]. These studies generated 5,177 microbial taxonomic profiles from 16S rRNA genes and over 3.5 terabases of metagenomic sequences. In gut samples, the Human Microbiome Project recovered more nonredundant microbial genes than MetaHIT, thereby expanding the gut community gene catalogue to more than 5 million genes.

Functional screening by shotgun whole genome analysis relies on sequencing of all the genetic material in the microbial community, including taxonomically unknown members, and matching the sequences to known functional genes. Such studies have generated fascinating information about functions within the microbial communities of the human gut. The extensive nonredundant catalogue of the microbial genes in the intestinal tract encodes groups of proteins engaged in up to 20,000 biological functions related with life in the intestinal habitat [4]. Some functions are common to free-living bacteria, like the main metabolic pathways (e.g. central carbon metabolism, amino acid synthesis), and some important protein complexes (RNA and DNA polymerases, ATP synthase, general secretory apparatus). Some other gene clusters encode functions which may be especially important for microbial life within the gut, such as those involved in adhesion to host proteins (collagen, fibrinogen, fibronectin) or in harvesting sugars from the glycolipids secreted by epithelial cells [4].

Interestingly, despite the highly divergent compositions of gut microbiota across individuals in terms of taxonomy, functional gene profiles are rather similar in healthy subjects. In contrast to taxonomic diversity, most functional pathways are common and expressed in similar abundance among fecal microbiotas from different human individuals [7]. Thus, such data imply that there is functional redundancy across taxonomic diversity, i.e. the same or similar functional pathways are present in different microbial species. This concept is likely to be very relevant for a definition of a 'nor-

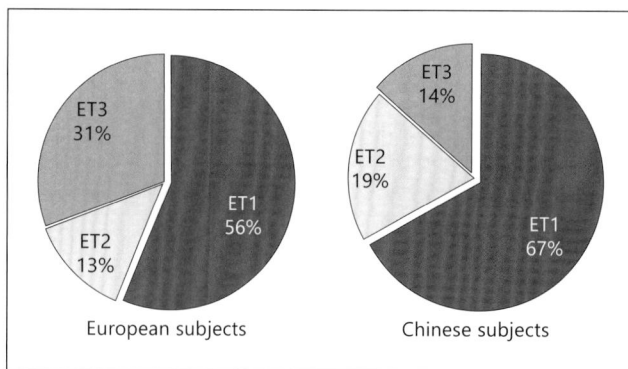

Fig. 2. Enterotype distribution of gut microbiotas in subjects from Europe and China. Data obtained from supplementary files in the studies by Arumugam et al. [5] and Qin et al. [10].

mal' or 'healthy' gut microbial ecosystem in humans: functional profiling may eventually become the optimal approach rather than merely the listing of species or strains present in the habitat.

Enterotypes

Network analysis of species abundance across different individuals suggested that the human microbiome is represented by well-balanced host-microbial symbiotic states driven by groups of co-occurring species and genera [5]. This hypothesis was investigated using a dataset of metagenomic sequences from American, European, and Japanese individuals. Multidimensional cluster analysis and principal component analysis revealed that all individual samples formed three robust clusters, which were designated as 'enterotypes' [5]. Each of the three enterotypes was identifiable by variations in the levels of one of three genera: *Bacteroides* (enterotype 1), *Prevotella* (enterotype 2), and *Ruminococcus* (enterotype 3). The basis for the enterotype clustering is unknown, but appears independent of nationality, sex, age, or BMI. As shown in figure 2, a recent study identified the three enterotypes in the Chinese population [10]. The enterotype concept suggests that enteric microbiota variations across individuals are generally stratified, not continuous.

In a cohort of US citizens, the reported enterotype partitioning was found to be related to long-term dietary patterns [11]. The *Bacteroides* enterotype was associated with diets enriched in protein and fat. In contrast, the *Prevotella* enterotype was linked to diets with a predominance of carbohydrates and sugars. The study in healthy children and adults from the Amazon region of Venezuela, Malawi, and the USA also identified the enterotypes' pattern when analyzing adults from the USA [9]. However, when including children from the USA and children and adults from developing countries, the enterotype picture got blurred for variation driven in adults by a trade-off between *Prevotella* and *Bacteroides*. In most adults from Amazonas and Malawi,

Prevotella is a dominant genus, whereas *Bacteroides* is not. This finding is consistent with the long-term dietary patterns previously identified as linked to the *Prevotella* enterotype. The original study by Arumugam et al. [5] also revealed that the microbiotas of children would not settle into the enterotype distribution.

Dysbiosis

Pathologies such as inflammatory bowel diseases [12], obesity [13, 14], type 2 diabetes [10], irritable bowel syndrome [15], *Clostridium difficile*-associated disease, and others [15] have been linked with changes in the composition of the gut microbiota. Consistency among studies is still poor for some of these examples. In addition, such associations do not necessarily indicate a causative role of the microbiota in the pathogenesis of disease, as they could be a consequence of the disease. Follow-up studies and, particularly, intervention studies aiming at restoring the normal composition of the gut microbiota will be needed.

It is important to emphasize that therapeutic attempts to restore the microbial disbalance underlying disease necessarily implicates an ecological approach. In this way, resilience, defined as the ability of an ecosystem to return to its original state after being disturbed, needs to be taken into account. Persistent gut colonization by infusion of either a complex foreign microbiota or a consortium of microbes rather than single strains seems to be feasible [16], as demonstrated in animal models [17]. This is also exemplified by fecal microbiota transplantation for treatment of *C. difficile*-associated disease [18]. It has been recently shown that infusion of allogenic intestinal microbiota from human lean donors to male recipients with metabolic syndrome, randomly controlled with autologous microbiota infusion, improved insulin sensitivity along with levels of butyrate-producing intestinal microbes [19]. The clinical implications of such changes need further investigations, but they clearly indicate that the manipulation of the gut microbiota is emerging as a therapeutic procedure.

Conclusions

New sequencing technologies together with the development of biocomputer analysis tools allow a description of the full spectrum of the microbial communities that inhabit the human intestinal tract, as well as their functional contributions to host health. The next steps include the identification of changes that are associated with, and may play a role in, certain disease states. A better knowledge of the contributions of microbial symbionts to host health and disease will certainly help in the design of new potential interventions to improve symbiosis and combat disease.

References

1 Handelsman J, Rondon MR, Brady SF, et al: Molecular biological access to the chemistry of unknown soil microbes: a new frontier for natural products. Chem Biol 1998;5:R245–R249.

2 Frank DN, Pace NR: Gastrointestinal microbiology enters the metagenomics era. Curr Opin Gastroenterol 2008;24:4–10.

3 Eckburg PB, Bik EM, Bernstein CN, et al: Diversity of the human intestinal microbial flora. Science 2005;308:1635–1638.

4 Qin J, Li R, Raes J, et al, MetaHIT Consortium, Bork P, Ehrlich SD, Wang J: A human gut microbial gene catalogue established by metagenomic sequencing. Nature 2010;464:59–65.

5 Arumugam M, Raes J, Pelletier E, et al, MetaHIT Consortium: Enterotypes of the human gut microbiome. Nature 2011;473:174–180.

6 Human Microbiome Project Consortium: Structure, function and diversity of the healthy human microbiome. Nature 2012;486:207–214.

7 Human Microbiome Project Consortium: A framework for human microbiome research. Nature 2012; 486:215–221.

8 Caporaso JG, Lauber CL, Costello EK, et al: Moving pictures of the human microbiome. Genome Biol 2011;12:R50.

9 Yatsunenko T, et al: Human gut microbiome viewed across age and geography. Nature 2012;486:222–227.

10 Qin J, Li Y, Cai Z, et al: A metagenome-wide association study of gut microbiota in type 2 diabetes. Nature 2012;490:55–60.

11 Wu GD, Chen J, Hoffmann C, et al: Linking long-term dietary patterns with gut microbial enterotypes. Science 2011;334:105–108.

12 Manichanh C, Borruel N, Casellas F, Guarner F: The gut microbiota in IBD. Nat Rev Gastroenterol Hepatol 2012;9:599–608.

13 Ley RE, Turnbaugh PJ, Klein S, Gordon JI, et al: Microbial ecology: human gut microbes associated with obesity. Nature 2006;444:1022–1023.

14 Tremaroli V, Backhed F: Functional interactions between the gut microbiota and host metabolism. Nature 2012;489:242–249.

15 Cho I, Blaser MJ: The human microbiome: at the interface of health and disease. Nat Rev Genet 2012;13: 260–270.

16 Lozupone CA, Stombaugh JI, Gordon JI, Jansson JK, Knight R: Diversity, stability and resilience of the human gut microbiota. Nature 2012;489:220–230.

17 Manichanh C, Reeder J, Gibert P, Varela E, Llopis M, Antolin M, Guigo R, Knight R, Guarner F: Reshaping the gut microbiome with bacterial transplantation and antibiotic intake. Genome Res 2010;20: 1411–1419.

18 Khoruts A, Dicksved J, Jansson JK, Sadowsky MJ: Changes in the composition of the human fecal microbiome after bacteriotherapy for recurrent *Clostridium difficile*-associated diarrhea. J Clin Gastroenterol 2010;44:354–360.

19 Vrieze A, et al: Transfer of intestinal microbiota from lean donors increases insulin sensitivity in individuals with metabolic syndrome. Gastroenterology 2012;143:913–916.

Dr. Virginia Robles Alonso
Digestive System Research Unit, University Hospital Vall d'Hebron
Passeig Vall d'Hebron, 119-129
ES–08035 Barcelona (Spain)
E-Mail vrobles@vhebron.net

Guarino A, Quigley EMM, Walker WA (eds): Probiotic Bacteria and Their Effect on Human Health and Well-Being.
World Rev Nutr Diet. Basel, Karger, 2013, vol 107, pp 25–31 (DOI: 10.1159/000345748)

The Intestinal Microbiota and Aging

Paul W. O'Toole[a] · Patrizia Brigidi[b]

[a]Department of Microbiology and Alimentary Pharmabiotic Centre, University College Cork, Cork, Ireland;
[b]Department of Pharmaceutical Science, University of Bologna, Bologna, Italy

Abstract

The intestinal microbiota of humans and other animals is an important determinant of health because it has properties that program the innate immune system, affect the availability of nutrients and bioactives from the diet, regulate barrier function, and restrict pathogen access to the intestinal epithelium. These properties are even more important when the host animal advances into older age; paradoxically, however, this is a life-stage when the microbiota appears to be in a state of flux. This review will describe how the application of molecular methods rather than culture-based approaches has recently yielded new insights into the changes in the human intestinal microbiota upon aging. At the extreme end of older life – in centenarians – even more pronounced changes in the intestinal microbiota occur. One of the most important drivers of changes in the microbiota is diet, and lack of diversity in the diet upon aging is linked to a less diverse microbiota, which correlates with poorer health status. We conclude by reviewing prospects for enhancing the health of older persons by a combination of microbiota profiling and microbiota adjustment.

The global proportion of people aged over 65 years is increasing dramatically. Less than 10 years from now, the global proportion of people aged over 65 years will exceed the proportion of children aged younger than 5 years for the first time in recorded history, and people in the over-65 age bracket will increase in number by 870,000 per month [1]. In Europe, citizens over the age of 65 years will increase from an overall proportion of 17.4% in 2010 to an estimated 29.5% by 2060 [2]. In the context of promoting the health of these citizens, the intestinal microbiota is an obvious target. In less than 10 years, the human intestinal microbiota has gone from being an unfashionable topic in microbial ecology to the center stage of health-related microbiological research [3]. Large catalogues of microbial taxa and their genes have been catalogued in North American [4, 5] and European consortia [6, 7]. During this period, correlations have been established in animal models or in humans between alterations of the microbiota and many human traits or syndromes (reviewed elsewhere in this volume).

At the two extremes of life – infancy and old age, the intestinal microbiota is in a state of flux (reviewed in [8]). In the case of infants, as reviewed elsewhere in this volume, the cause of microbiota variability is the transition from a near-sterile state in utero to the microbe-laden external world. In older adults, a generally opposite trend was indicated by culture-based studies nearly four decades ago [9] involving changes at the phylum and genus level, but reproducible taxon alterations and underlying reasons were not clear. This review describes recent advances in understanding how and why the intestinal microbiota of older people differs from that of older adults and discusses the significance for monitoring and maintaining health.

Physiology of Aging and Its Impact on the Microbiota

The changing physiology associated with aging presents a different physical environment for gut bacteria (reviewed in [8]). Older subjects may experience problems with dentition and chewing strength, which can alter their diet selection patterns and the particle size and digestibility of food entering the alimentary tract. There may be problems with saliva production, which affects starch breakdown and has an impact on diet choices. The transit time for digesta increases with age due to reduced muscle tone and intestinal motility [10], which can also affect diet choices (i.e. avoidance of fiber or necessity for laxatives). Thus, there are intertwined physiological and dietary factors that can have an impact on the microbiota. Another characteristic feature of aging is persistent inflammation and immunosenescence, termed 'inflammaging', which involves persistent NF-κB-regulated innate immune activity, and a loss of naïve CD4+ T cells [11]. Infectious disease is an overlooked cause of mortality in older persons, and an altered intestinal microbiota is a plausible environmental factor in propagating inflammaging or in loss of colonization resistance leading to increased susceptibility to intestinal infection [12].

The Microbiota of Older Persons: Culture-Based Analyses

In almost all the studies reviewed here, feces was used as a surrogate for the distal gut because of the simplicity of sampling. Based upon the earliest culture-based analyses, the gut microbiota of older persons was marked by a proportional reduction in bifidobacteria numbers compared to younger adults, and by an increase in the relative proportions of coliforms, enterococci, and selected *Clostridium* species [9]. A later study reported changes in phylum Bacteroidetes genus proportions, and confirmed lower abundance of bifidobacteria [13]. After an intervening period punctuated by studies which often conflicted, Woodmansey et al. [14] employed a combination of culture-based and chemical analyses to identify a general decrease in species diversity for *Bacteroides*, *Prevotella*, *Bifidobacterium*, and *Lactobacillus* spp., and proportional enrichment of Enterobacteriaceae, staphylococci, streptococci, and *Candida albicans*, compared to young adults. Culture-dependent analyses were summed up (just before the widespread application of molecular studies) in a review which highlighted reduction in bifidobacteria and *Bacteroides* diversity and increased fusobacteria, clostridia, and eubacteria levels in the intestinal microbiota of older subjects [15].

Recent Insights from Culture-Independent Analyses

In one of the earliest culture-independent analyses, Hayashi et al. [16], using a 16S rRNA gene clone library sequencing approach, showed that the intestinal microbiota of seniors had an unusually high proportion of novel phylotypes. Costs at that time restricted this kind of analysis to 6 subjects, but qPCR allowed a larger cohort to be analyzed, which revealed reductions in proportions of the *Bacteroides-Prevotella* group, in bifidobacteria, in *Desulfovibrio* spp., some clostridia, and in *Faecalibacterium prausnitzii* upon hospitalization [17]. Van Tongeren et al. [18] used hybridization with group-specific probes to demonstrate interesting microbiota changes associated with frailty, including a significant reduction in the number of lactobacilli (26-fold), the *Bacteroides-Prevotella* group (threefold) and *F. prausnitzii* (fourfold), whereas the number of Enterobacteriaceae was significantly higher (sevenfold) in fecal samples from very frail volunteers. Mueller et al. [19] reported inconsistency in trends in aging-related microbiota composition alterations across four European countries, pointing to the likely impact of environmental factors.

To address problems associated with low subject numbers and single residence locations that typified some previous studies, the ELDERMET project (http://elder-met.ucc.ie) examined the fecal microbiota in 500 subjects in various residence locations, while also recording detailed clinical, anthropometric, and cognitive measurements for each participant. The baseline cross-sectional microbiota analysis of the first 161 subjects (mean age 78 ± 7 years) revealed a core microbiota that differed from that of healthy younger adults [20]. The aggregate microbiota (i.e. averaged by composition across the cohort) suggested an unusual dominance of the phylum *Bacteroidetes* (57% by proportion). However, later analysis (see below) showed that this is heavily influenced by the proportions of subjects of different health states and diet types. Each subject had a distinct microbiota composition pattern when viewed at sufficient phylogenetic depth. There was a striking degree of variation of the phylum composition across the cohort, with huge differences in the proportions of the phylum *Bacteroidetes* and phylum *Firmicutes* [20], but reasons for microbiota variation, or correlations with the metadata, were not examined in this baseline analysis.

Extreme Aging: The Microbiota of Centenarians

Centenarians can be considered the best example of healthy aging in humans since they live 20–30 years longer than the other members of their demographic cohort and they die in a few days. Furthermore, centenarians avoid or delay the major inflammation-driven age-related diseases, such as cardiovascular diseases, diabetes mellitus, Alzheimer's disease, frailty, and cancer [21, 22]. Thus, they represent a valuable model for studying how the reshaping of the gut microbiota impacts on the progression of immunosenescence and inflammaging.

Recently, Biagi et al. [23] performed a comparative study to characterize the gut microbiota composition in different age groups, including young adults (mean age 30 years), elderly (mean age 70 years), and centenarians (mean age 100 years, living in their own household, and showing good physical and cognitive health conditions). All the 84 subjects analyzed were recruited from a restricted geographic area of Northern Italy. The fecal microbiota was characterized using the HITChip microarray, which includes 16S rRNA probes specific for 1,140 bacterial phylotypes [24].

Hierarchical clustering of the HITChip profiles showed that centenarians tended to group together, whereas no separation was observed with the subjects belonging to the elderly and young groups. Centenarians showed a significantly compromised intestinal microbiota, whereas the microbiota of the elderly group was unexpectedly similar to that of the younger adults. In particular, the fecal microbiota of centenarians was enriched in many opportunistic, and possibly pathogenic, bacterial groups belonging to Proteobacteria and Bacilli. It was also depleted in many butyrate producers belonging to *Clostridium* cluster XIVa and characterized by a rearrangement in the composition represented by *Clostridium* cluster IV, with a decrease of *F. prausnitzii* and an increase of the *C. leptum* group. These results suggest that changes in gut microbiota do not follow a linear relation with age, and that the compositional stability of the adult gut microbiota could last longer than expected. Furthermore, centenarians showed an inflammation score significantly higher than in the other age groups, with increased plasma levels of IL-6 and IL-8. The dysbiosis of the gut microbiota in centenarians could be related to the proinflammatory cytokine pattern, with several Proteobacteria species which positively correlated with IL-6 and IL-8 and *Clostridium* cluster XIVa groups which inversely correlated with IL-6 and lL-8. These findings confirm that the age-related changes in the gut microbiota may concord with the complex process that both sustain and is promoted by the overall inflammatory process typical of advancing age.

Diet-Microbiota-Health Relationships

In an attempt to rationalize the microbiota composition variation previously noted in older subjects [20], and to examine correlations with the metadata, the ELDERMET consortium recently reported an analysis of relationships between microbiota, residence location, diet, and health status in 173 adults (78 ± 8 years) who had dietary data available and had not been treated with antibiotics. Surprisingly, the microbiota composition was strongly linked to where people resided in the community, with a major split in the data depending on whether subjects resided at home or in long-term residential care. The dietary intake data defined distinct groups along the same residence lines (perhaps not surprisingly), but detailed bioinformatics and statistical analysis showed significant clustering between the diet and microbiota clusterings, and the change of diet in moving from the community to long-term residential care appeared to be responsible for altered microbiota [25]. A healthy diverse diet in community-dwelling subjects was associated with a diverse microbiota, whereas a lower diversity diet in long-term residence was associated with a significant reduction in microbiota diversity. Metabolomics and shotgun microbiome sequencing respectively confirmed higher levels of certain short-chain fatty acids in fecal water samples and the corresponding microbial genes for their production in community-dwelling compared to long-term care subjects [25]. Comparison of gradients in microbiota composition with those in clinical parameters revealed correlations with indicators of frailty, cognitive function, and inflammation; however, these need to be confirmed in prospective studies.

Future Prospects for Research in the Microbiota in Older Persons

Based upon its variability and its health connections, the intestinal microbiota of older persons represents a plausible target for increasing lifespan and quality of life in old age [26]. A simple microbiota modulation point would appear to be diet, but this will require intelligent solutions for product size, formulation, and flavoring to overcome the physiological challenges in older subjects mentioned above. Dietary supplements must also be affordable to the consumer. Nu-Age (http://www.nu-age.eu/) is a European Union Framework Programme 7 project (of which the authors are partners in), which aims to improve the health and quality of life in the European Union's ageing population by counteracting inflammaging through a diet approach. One of the assumptions of Nu-Age is that diet can modulate the composition and functionality of the gut microbiota, inducing systemic responses. A large number of subjects (65–79 years of age) are being recruited and characterized before and after a specific dietary intervention by measuring selected robust parameters (nutritional status; physical and cognitive functions; immunological, biochemical, and metabolic parameters; genetics and epigenetics) as well as by analyzing the gut microbiome using advanced 'omics' techniques. All the data will be integrated adopting a systems biology approach to fill the current lack of knowledge on how diet can impact and counteract age-related decline.

In addition to diet-based approaches, improved understanding of the relevant microbiota components whose proportions change significantly in less healthy older people may also allow their direct therapeutic replacement, akin to the next generation of probiotics. There is already enough information to implement microbiota profiling to identify subjects with the less diverse microbiota composition associated with reduced health. Microbiota profiling and surveillance, combined with microbiota modulation by altered diet or therapeutic intervention, offer the prospect of decelerating microbiota-related health reduction upon aging.

Acknowledgements

Work in the authors' laboratories is supported by awards from Science Foundation Ireland, the Health Research Board Ireland, the Department of Agriculture Food and Marine of the Irish Government, the Italian Ministry of University and Research (MiUR), and the European Union FP7 award to Nu-Age.

References

1 Kinsella K, He W: An Aging World: 2008. International Population Reports. Washington, US Census Bureau, 2009.
2 European Commission: Population Structure and Ageing. 2011. http://epp.eurostat.ec.europa.eu/statistics_explained/index.php/Population_structure_and_ageing.
3 Gordon JI: Honor thy gut symbionts redux. Science 2012;336:1251–1253.
4 The Human Microbiome Project Consortium: Structure, function and diversity of the healthy human microbiome. Nature 2012;486:207–214.
5 The Human Microbiome Consortium: A framework for human microbiome research. Nature 2012;486:215–221.
6 Arumugam M, et al: Enterotypes of the human gut microbiome. Nature 2011;473:174–180.
7 Qin J, et al: A human gut microbial gene catalogue established by metagenomic sequencing. Nature 2010;464:59–65.
8 O'Toole PW, Claesson MJ: Gut microbiota: changes throughout the lifespan from infancy to elderly. Int Dairy J 2010;20:281–291.
9 Mitsuoka T: Intestinal Bacteria and Health. Tokyo, Harcourt Brace Jovanovich, 1978.
10 Camilleri M, et al: Insights into the pathophysiology and mechanisms of constipation, irritable bowel syndrome, and diverticulosis in older people. J Am Geriatr Soc 2000;48:1142–1150.
11 Franceschi C, et al: Inflamm-aging. An evolutionary perspective on immunosenescence. Ann NY Acad Sci 2000;908:244–254.
12 Guigoz Y, Dore J, Schiffrin EJ: The inflammatory status of old age can be nurtured from the intestinal environment. Curr Opin Clin Nutr Metab Care 2008;11:13–20.
13 Hopkins MJ, Macfarlane GT: Changes in predominant bacterial populations in human faeces with age and with *Clostridium difficile* infection. J Med Microbiol 2002;51:448–454.
14 Woodmansey EJ, et al: Comparison of compositions and metabolic activities of fecal microbiotas in young adults and in antibiotic-treated and non-antibiotic-treated elderly subjects. Appl Environ Microbiol 2004;70:6113–6122.
15 Woodmansey EJ: Intestinal bacteria and ageing. J Appl Microbiol 2007;102:1178–1186.
16 Hayashi H, et al: Molecular analysis of fecal microbiota in elderly individuals using 16S rDNA library and T-RFLP. Microbiol Immunol 2003;47:557–570.

17 Bartosch S, et al: Characterization of bacterial communities in feces from healthy elderly volunteers and hospitalized elderly patients by using real-time PCR and effects of antibiotic treatment on the fecal microbiota. Appl Environ Microbiol 2004;70:3575–3581.

18 van Tongeren SP, et al: Fecal microbiota composition and frailty. Appl Environ Microbiol 2005;71:6438–6442.

19 Mueller S, et al: Differences in fecal microbiota in different European study populations in relation to age, gender, and country: a cross-sectional study. Appl Environ Microbiol 2006;72:1027–1033.

20 Claesson MJ, et al: Composition, variability, and temporal stability of the intestinal microbiota of the elderly. Proc Natl Acad Sci USA 2011;108(suppl 1):4586–4591.

21 Motta M, et al: Diabetes mellitus in the extreme longevity. Exp Gerontol 2008;43:102–105.

22 Salvioli S, et al: Why do centenarians escape or postpone cancer? The role of IGF-1, inflammation and p53. Cancer Immunol Immunother 2009;58:1909–1917.

23 Biagi E, et al: Through ageing, and beyond: gut microbiota and inflammatory status in seniors and centenarians. PLoS One 2010;5:e10667.

24 Rajilic-Stojanovic M, et al: Development and application of the human intestinal tract chip, a phylogenetic microarray: analysis of universally conserved phylotypes in the abundant microbiota of young and elderly adults. Environ Microbiol 2009;11:1736–1751.

25 Claesson MJ, et al: Gut microbiota composition correlates with diet and health in the elderly. Nature 2012;488:178–185.

26 Ottaviani E, et al: Gut microbiota as a candidate for lifespan extension: an ecological/evolutionary perspective targeted on living organisms as metaorganisms. Biogerontology 2011;12:599–609.

Dr. Paul W. O'Toole
Room 447 Food Science Building
Department of Microbiology, University College Cork
Cork (Ireland)
E-Mail pwotoole@ucc.ie

Guarino A, Quigley EMM, Walker WA (eds): Probiotic Bacteria and Their Effect on Human Health and Well-Being.
World Rev Nutr Diet. Basel, Karger, 2013, vol 107, pp 32–42 (DOI: 10.1159/000346493)

Shaping Intestinal Bacterial Community by TLR and NLR Signaling

Koichi S. Kobayashi[a–c]

[a]Department of Microbial and Molecular Pathogenesis, Texas A&M Health Science Center, College Station, Tex., [b]Department of Microbiology and Immunobiology, Harvard Medical School, and [c]Department of Cancer Immunology and AIDS, Dana-Farber Cancer Institute, Boston, Mass., USA

Abstract

Human intestines harbor a diverse microbial community composed of a large number of bacteria and other microorganisms. These intestinal microbiota have evolved to achieve a symbiotic relationship with the host. In addition to aiding host metabolic pathways by breaking down foods and supplying nutrients to their host, microbiota play an important role in the development and maintenance of the host immune system. At the same time, the detection of microorganisms and their products by innate immune receptors such as TLRs (Toll-like receptors) and NLR (nucleotide-binding domain, leucine-rich repeats, or NOD-like receptor) proteins are critical for maintaining intestinal homeostasis by shaping microbial communities. This review summarizes recent progress about the role of TLRs and NLRs in the regulation of intestinal microbiota. Accumulating evidence suggest that intestinal microbiota have a large impact on both intestinal and systemic diseases. Therefore, understanding the mechanism of microbial regulation by TLRs and NLRs is important for the advancement of therapeutic interventions against digestive and other diseases.

Copyright © 2013 S. Karger AG, Basel

The human intestine harbors microbial communities that are large and diverse in number, with estimations at 10^{13}–10^{14} bacteria per individual. The composition and concentration of intestinal microbiota are regulated by a multitude of factors such as genetic background, diet, and interaction between commensal and pathogenic bacteria [1]. Major bacterial taxa in the human microbiota belong to the Firmicutes and Bacteroidetes phyla, whereas minor populations include the Proteobacteria, Acinobacteria, Verrucomicrobia, or Fusobacteria phyla [2]. The small intestine harbors significantly less bacteria compared to the large intestine. Also, the composition of bacterial communities between the small and large intestines differ significantly [3]. This intestinal ecosystem has evolved to allow both the host and microbiota to benefit from their balanced symbiotic relationship [4]. The microbiota contributes to host nutrition and energy balance as well as to the development of the mucosal immune system. In return, the bacterial flora is assured of a stable habitat in the host intestine. Although the host-microbiota symbiotic relationship has coevolved to be beneficial for both parties, a large array of intestinal microorganisms can be a potential threat to the host. Therefore, outgrowth of pathogenic bacteria or the opportunistic invasion of the epithelial barrier by resident bacteria should be minimized. Consequently, the human intestinal immune system has evolved to protect host tissues from microorganisms and simultaneously maintain the symbiotic benefits from the microbial presence [5]. The intestinal microbiota contributes to the development of the immune system and enhances resistance against infection by pathogenic bacteria. At the same time, the intestinal mucosal immunity can shape bacterial communities by sensing the presence of commensal and pathogenic bacterial via pattern recognition receptors, such as TLRs (Toll-like receptors) and NLR (nucleotide-binding domain, leucine-rich repeats, or NOD-like receptor) proteins. Here I summarize recent advancement of this field, which is critical to understanding intestinal as well as systemic inflammatory diseases.

Role of TLR Signaling in the Regulation of Luminal Bacterial Communities at a Steady State

Since sensing bacteria by TLRs elicit strong antimicrobial immune responses, it was interesting to see if TLR signaling would change the composition of intestinal bacterial flora at a steady state. Recently, the extent to which TLRs influence the composition of the intestinal microbiota was investigated by the 16s ribosomal RNA gene sequencing of samples taken from ilea, cecum, and feces of TLR- and MyD88-deficient mice [6]. Analysis of pyrosequencing data obtained from MyD88-, TLR2-, TLR4-, TLR5-, and TLR9-deficient mice demonstrated that the impact of TLR deficiency on the composition of the intestinal microbiota is surprisingly minimal under homeostatic conditions [6]. This observation was further illustrated by similar microbiota composition among strains after recovery from antibiotic treatment, further confirming a minimal impact of TLRs on the microbiota composition [6]. Interestingly, the microbiota of MyD88- and TLR-deficient mice are markedly different. A similar study demonstrated that MyD88-deficient mice harbor a significantly increased level of Rikenellaceae and Porphyromonadaceae bacterial families in their cecal contents [6]. This may suggest that deficiency of a single TLR may not have a significant impact on the intestinal bacterial community. Alternatively, this may indicate that the IL-1/IL-18 pathway, in addition to the TLR signaling pathway, may play an important role in the regulation of microbiota. This is reminiscent of the phenotype of IL-18-, ASC-, or Nlrp6-deficient mice, which will be discussed in the NLR section of this review.

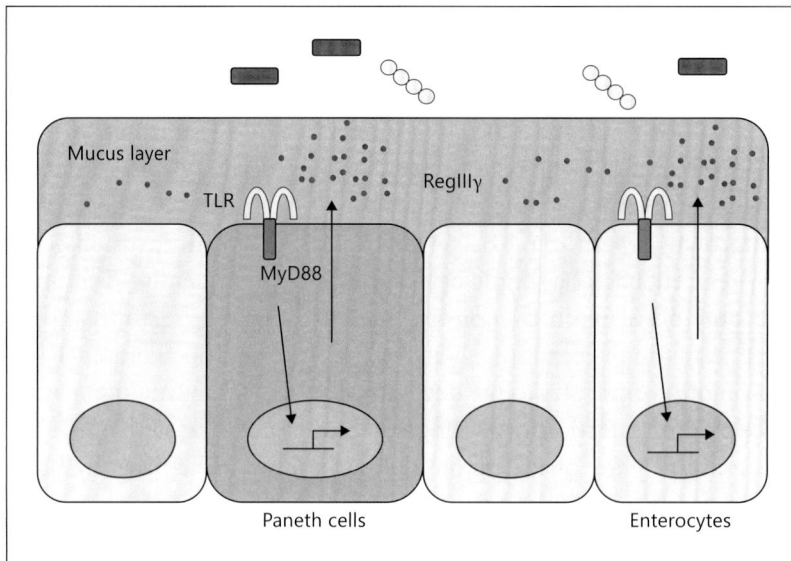

Fig. 1. RegIIIγ-dependent physical barrier in the small intestine. The human intestine has developed multiple strategies to protect the intestinal epithelial cells from invasion by resident and pathogenic bacteria. Enterocytes, the most abundant cell type in the intestinal epithelium, secrete antimicrobial proteins such as RegIIIγ. Paneth cells, which are specialized epithelial cells in the ileal crypts, produce abundant antimicrobial compounds including α-defensins and RegIIIγ to regulate bacterial populations. The expression of RegIIIγ is induced by bacterial products through TLR sensing, which is dependent on the MyD88 adaptor molecule. Goblet cells produce mucin glycoproteins, constituents of the mucin mucus layer. RegIIIγ is enriched in the mucus layer that physically separates the microbiota from the small intestinal epithelial surface.

TLR-Dependent Microbiota Regulation through the Secretion of RegIIIγ

Antimicrobial peptides of the REG3 family (such as RegIIIγ in mice or REG3α in humans) are C-type lectins expressed in the small intestines of mice and humans and in the large intestine during intestinal infection or inflammation [7–9]. RegIIIγ displays antibacterial activity against Gram-positive bacteria by targeting their bacterial cell wall peptidoglycan [10]. Induction of RegIIIγ is TLR signaling-dependent since the expression of RegIIIγ is induced by stimulation with TLR ligands such as lipopolysaccharide or flagellin (ligands for TLR4 and TLR5, respectively) [11–13], and the RegIIIγ transcript is low in germ-free mice or MyD88-deficient mice (fig. 1) [9, 11]. Strikingly, the expression of RegIIIγ is required for maintaining an approximate 50-μm zone that physically separates the microbiota from the small intestinal epithelial surface (fig. 1) [14]. This separation is important to restrict the number of bacteria attached to the epithelial layers since the number of mucosa-associated, but not luminal bacteria, are significantly increased in both RegIIIγ-deficient mice and MyD88-deficient mice [14]. Furthermore, Paneth cell-specific MyD88 signaling is critical in limiting penetration of both commensal and pathogenic bacteria into host tissues through

RegIIIγ secretion [11]. Consistent with this is the fact that RegIIIγ induction though MyD88 protects the mice from infection of *Listeria monocytogenes* and vancomycin-resistant *Enterococcus faecium* [12, 15]. Therefore, epithelial cells (particularly Paneth cells) can directly sense an intestinal bacterial community through TLRs and maintain homeostasis at the intestinal host-microbial interface via production of antibacterial RegIIIγ.

Establishments of Commensal Colonization via TLR2 Signaling

The role of mucosal immune systems to protect host tissues against intestinal bacterial flora has long been appreciated and is considered one of its major functions. The opposite case, however, may be possible upon TLR activation. It has recently been reported that TLR2 can aid the colonization of commensal bacteria, *Bacteroides fragilis* [16]. *B. fragilis* carries a symbiotic factor, polysaccharide A (PSA) which displays immunomodulatory activity and suppresses intestinal inflammation [17, 18]. PSA shapes host immune responses by inducing IL-10 and IFN-γ from CD4 T cells through TLR2 signaling [16, 18]. Monocolonization of mutant *B. fragilis* lacking PSA into germ-free mice resulted in poor colonization, which could be rescued by coadministrating purified PSA [16]. PSA-dependent colonization of *B. fragilis* is associated with an increase of T regulatory cells and a decrease of Th17 cell responses. Indeed, colonization of PSA-deficient *B. fragilis* can be augmented by the administration of anti-IL-17 antibody [16]. Therefore, *B. fragilis* colonizes the intestine efficiently by inducing an immune tolerant state through the activation of TLR2 signaling. This study presents an interesting case in terms of how bacteria may utilize the host immune system for their own benefit. There are many issues still to be addressed. The function of PSA is not clear, as other TLR2 ligands cannot be replaced with PSA for the induction of T regulatory cells. Also, the study was mostly performed by the monoassociation of *B. fragilis* into germ-free mice. Therefore, it is still possible that PSA may affect the immune system differently if other bacterial strains coexist in the intestine. However, these studies demonstrated that the outcome of innate immune activation may not simply be an elimination of bacteria but can be an enhancement of symbiosis of bacteria with host, depending on bacterial strains and colonization conditions.

Role of TLR5 in the Regulation of Intestinal Microbiota and Homeostasis

TLR5 recognizes flagellin, a major component of bacterial flagella that is required for bacterial motility [19]. Studies using mice deficient in TLR5 indicate that TLR5 may play a role in the regulation of bacterial communities in the intestine. The original TLR5-deficient mice generated by multiple groups displayed no obvious phenotype unless experimentally challenged [20, 21]. Intriguingly, it was reported that the same

TLR5-deficient mice developed spontaneous colitis when kept in a different animal facility, indicating the critical role of TLR5 in intestinal homeostasis [22]. TLR5-deficient mice also developed a phenotype similar to a metabolic syndrome characterized with obesity, increased fat-pad mass, higher blood pressure, and reduced glucose tolerance [23]. This is due to altered microbiota in TLR5-deficient mice as the transfer of TLR5 microbiota into germ-free mice resulted in the development of a similar metabolic syndrome phenotype [23]. The colitogenic microbiota in TLR5-deficient mice seems specific to this facility since no spontaneous colitis or metabolic syndrome phenotype was found in other animal facilities [24]. The association of microbiota and colitis in TLR5-deficient mice was further confirmed by the pyrosequencing of bacterial community in TLR5-deficient mice [25]. Among littermates in TLR5-deficient mice, only a minor population developed colitis with the majority not showing symptoms. Microbiota of colitic TLR5-deficient mice, compared with noncolitic TLR5-deficient mice or wild-type mice, are characterized by transient high levels of proteobacteria, particularly enterobacteria species including *Escherichia coli*, in close proximity to the intestinal epithelium [25]. These studies indicate that TLR5 signaling is important for stabilizing gut microbiota and that bacterial dysbiosis may trigger or exacerbate pathogenic conditions such as inflammatory bowel diseases or metabolic syndromes. Also, the failure to reproduce the phenotype of TLR5-deficient mice in other facilities indicates that microbiota in each animal facility may significantly differ and those differences may greatly affect the pathogenic condition caused by an altered innate immune system.

NLR-Mediated Microbiota Modulation

The NLR family of proteins is another important arm of innate immunity that senses presence of microbes. NLR proteins consist of 22 members in humans. Most members are cytoplasmic proteins expressed in a wide variety of cells, including immune and epithelial cells [26]. NLR proteins play an important role in the recognition and defense against pathogens or extracellular danger signals [27, 28]. While Nod1 and Nod2 are involved in the activation of inflammatory gene expression, several NLRs are involved in the activation of caspase-1-activating complexes called inflammasomes, a multiprotein complex consisting of NLRs, caspase-1, and the adaptor protein ASC [28, 29]. These NLRs respond to various microbial products or damage-associated molecular patterns and contribute to the host defense by the maturation and release of the IL-1 family of inflammatory cytokines including IL-1β and IL-18 as well as the induction of programmed cell death called pyroptosis [28, 30, 31].

Nod2 is an important player in the intestine, as evidenced by its association with Crohn's disease, a major form of inflammatory bowel disease. Nod2 is highly expressed in dendritic cells [32], macrophages [33], and intestinal, lung, and oral epithelial cells [34–36], particularly in Paneth cells [37] and to a lesser extent in T cells [38, 39]. MDP (muramyl dipeptide), found in both the Gram-positive and Gram-

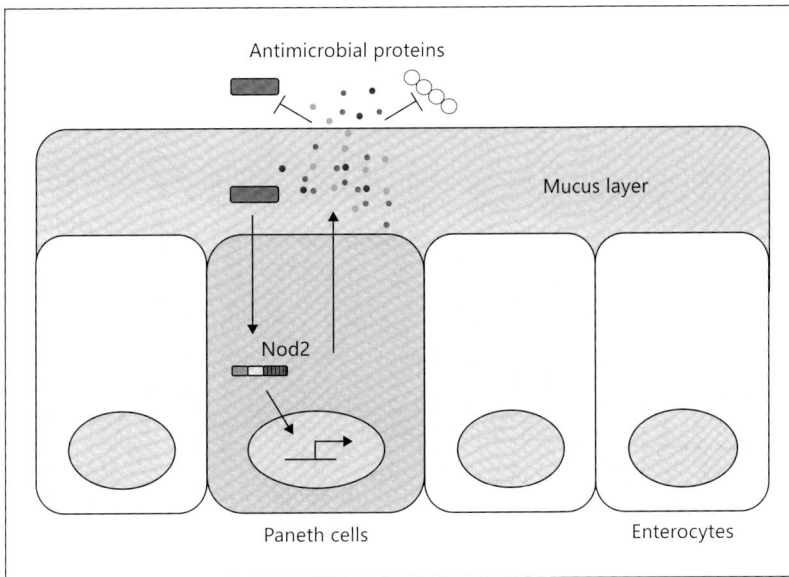

Fig. 2. Nod2-mediated regulation of bacteria by ileal Paneth cells. In the small intestine, Paneth cells sense bacteria or bacterial products and release antimicrobial peptides, which tightly regulate the bacterial flora. Nod2 is essential for bacteriocidal activity of Paneth cells.

negative bacteria cell wall [40, 41], was recognized by Nod2 and thereby provides resistance to a wide variety of bacteria. Nod2 plays a crucial role in regulating host-microbial interaction and maintaining intestinal homeostasis [42–44]. Nod2 is an important regulator of ileal crypt antimicrobial function (fig. 2) [42]. Indeed, *Nod2*-deficient crypts are also unable to kill bacteria efficiently through the secretion of antibacterial compounds from Paneth cells. Studies using Nod2-deficient mice indicate that Nod2 is not only important for MDP sensing in Paneth cells, but also for a general effect on secretion and/or composition of antibacterial factors. In agreement with the important function of Nod2 for Paneth cells, it was discovered that Nod2 is one of the critical factors that effectively regulate the bacterial concentration or load in the intestine. *Nod2*-deficient mice carry a significantly higher amount of *Bacteroides*, Firmicutes, and *Bacillus* in the terminal ileum compared to their wild-type littermates [42, 43]. In addition to commensal bacteria, Nod2 also controls the load of pathogenic bacteria in the ileum [42]. Interestingly, the expression of Nod2 is induced by commensal bacterial flora. Germ-free mice have poor expression of Nod2 in the intestine in comparison to conventional specific pathogen-free mice. Mono-colonization of germ-free mice with a single probiotic bacterial strain, such as *Lactobacillus plantarum* and *E. coli* strain Nissle 1917, increased the expression of Nod2. Therefore, there is a feedback loop in which bacteria-mediated upregulation of Nod2 counteracts on the bacterial flora to keep them under control in the ileum. A breakdown in this balanced relationship can lead to dysbiosis, which is known to underlie the pathogenesis of Crohn's disease [3]. Mutations in NOD2 are strongly associated with ileal Crohn's disease susceptibility in North American and European populations. Although the exact mechanism by which Nod2 mutations contribute to the pathogenesis of Crohn's disease is still unclear, loss-of-function mutations in Nod2 have been suggested to alter host-microbial interactions through various mechanisms, one of them being altered antimicrobial activity of Paneth cells in the terminal ileum [45]. Ileal Crohn's disease patients showed reduced levels of the Paneth cell-derived human α-defensins HD-5 and 6 [46–48]. Similarly, *Nod2*-deficient mice have reduced mRNA expression of Paneth cell-derived α-defensins and were more susceptible to oral infection with *Listeria monocytogenes* [49]. Interestingly, *Helicobacter hepaticus* inoculation of *Nod2*-deficient mice leads to the development of granulomatous inflammation of the ileum, which shares various similarities with ileal Crohn's disease in human patients [50]. *Nod2*-deficient mice were protected from *H. hepaticus*-induced ileal inflammation by restoring the Paneth cell bacterial killing activity of the mice, supporting that Nod2 function in Paneth cells is critical in preventing intestinal inflammation [50]. Therefore, Nod2 is necessary for normal Paneth cell function which in turn plays a vital role in maintaining the normal microbial population and gut homeostasis [51].

NLRP6 is another important NLR protein in the intestine, yet its mechanism is distinct from that of Nod2. NLRP6 is preferentially expressed in the epithelial cells of various organs including the intestine [52, 53]. NLRP6 can associate with caspase-1 via an adapter ASC, thus activating the inflammasome to generate mature forms of IL-1β and IL-18. Recently, it has been reported that NLRP6-deficient mice carry altered microbiota [53]. In particular, nine genera belonging to four phyla (Firmicutes, Bacteroidetes, Proteobacteria, and TM7) were significantly altered. An unnamed genus in the Prevotellaceae was most significantly increased in NLRP6-deficient mice, followed by the phylum TM7 and the named genus *Prevotella* within Prevotellaceae. On the other hand, members of the genus *Lactobacillus* in the Firmicutes phylum were reduced. Electron microscopic studies disclosed aberrant colonization of crypts of Lieberkuhn with bacteria with morphologic features of Prevotellaceae. Interestingly, this altered microbiota is associated with a colitis-prone phenotype of NLRP6 mice. NLRP6-deficient mice are highly susceptible to colitis induced by dextran sodium sulfate [52–54]. This colitis-prone phenotype is transmissible to wild-type mice by cohousing, indicating that the altered microbiota in NLRP6-deficient mice can be transferred to other strains. Although caspase-1 can activate both IL-1β and IL-18, it was determined that IL-18 plays a critical role in the microbiota regulation downstream of NLRP6 since IL-18 but not IL-1β-deficient mice could transfer the colitis-prone phenotype to wild-type mice by cohousing. Using various mutant mice, it was demonstrated that NLRP6 regulates colitogenic microbiota by the NLRP6-ASC-caspase-1-IL-18 axis. Interestingly, the altered microbiota is also associated with another disease entity, nonalcoholic fatty liver disease (NAFLD). The mice cohoused with *Nlrp6*-deficient mice fed with a methionine-choline-deficient diet developed a severe

NAFLD-like pathology including hepatic steatosis, inflammatory cell infiltration, and fibrosis [55], providing further supportive evidence that changes of intestinal microbiota may affect susceptibility to systemic disorders.

Conclusion

Recent advancements in mucosal immunology and microbiome studies have uncovered the critical roles of TLR and NLR proteins in the intestinal tract. Interestingly, the regulatory mechanism of microbiota by TLR and NLR protein activation varies, and even outcomes can be very different depending on the condition as seen in the effect of PSA from *B. fragilis* on immune tolerance and symbiotic bacterial colonization. Although recent NIH-funded multicentered human microbiome studies have shown that personal habits and genetic makeup are major factors in causing variations of bacterial community among individuals, the understanding of how a particular microbiome is established under the influence of certain foods, host genetic factors, and immune systems still has a long way to go. Further studies on the function of the innate immune system in the intestine is pivotal for understanding human intestinal microbiota and associated intestinal and systemic inflammatory diseases.

Acknowledgements

This work was supported by grants from the NIH (K.S.K. R01DK074738) and the Broad Medical Research Program of the Eli and Edythe L. Broad Foundations (K.S.K.). K.S.K. is a recipient of the Investigator Award from the Cancer Research Institute and the Claudia Adams Barr Award. The author thanks Yuen-Joyce Liu and Amlan Biswas for proofreading.

References

1 Begue B, Dumant C, Bambou JC, Beaulieu JF, Chamaillard M, Hugot JP, Goulet O, Schmitz J, Philpott DJ, Cerf-Bensussan N, Ruemmele FM: Microbial induction of CARD15 expression in intestinal epithelial cells via Toll-like receptor 5 triggers an antibacterial response loop. J Cell Physiol 2006;209:241–252.

2 Eckburg PB, Bik EM, Bernstein CN, Purdom E, Dethlefsen L, Sargent M, Gill SR, Nelson KE, Relman DA: Diversity of the human intestinal microbial flora. Science 2005;308:1635–1638.

3 Sartor RB: Microbial influences in inflammatory bowel diseases. Gastroenterology 2008;134:577–594.

4 Hooper LV, Macpherson AJ: Immune adaptations that maintain homeostasis with the intestinal microbiota. Nat Rev Immunol 2010;10:159–169.

5 Hooper LV, Littman DR, Macpherson AJ: Interactions between the microbiota and the immune system. Science 2012;336:1268–1273.

6 Ubeda C, Lipuma L, Gobourne A, Viale A, Leiner I, Equinda M, Khanin R, Pamer EG: Familial transmission rather than defective innate immunity shapes the distinct intestinal microbiota of TLR-deficient mice. J Exp Med 2012;209:1445–1456.

7 Christa L, Carnot F, Simon MT, Levavasseur F, Stinnakre MG, Lasserre C, Thepot D, Clement B, Devinoy E, Brechot C: HIP/PAP is an adhesive protein expressed in hepatocarcinoma, normal Paneth, and pancreatic cells. Am J Physiol 1996;271:G993–G1002.

8 Ogawa H, Fukushima K, Naito H, Funayama Y, Unno M, Takahashi K, Kitayama T, Matsuno S, Ohtani H, Takasawa S, Okamoto H, Sasaki I: Increased expression of HIP/PAP and regenerating gene III in human inflammatory bowel disease and a murine bacterial reconstitution model. Inflamm Bowel Dis 2003;9:162–170.

9 Cash HL, Whitham CV, Behrendt CL, Hooper LV: Symbiotic bacteria direct expression of an intestinal bactericidal lectin. Science 2006;313:1126–1130.

10 Lehotzky RE, Partch CL, Mukherjee S, Cash HL, Goldman WE, Gardner KH, Hooper LV: Molecular basis for peptidoglycan recognition by a bactericidal lectin. Proc Natl Acad Sci USA 2010;107:7722–7727.

11 Vaishnava S, Behrendt CL, Ismail AS, Eckmann L, Hooper LV: Paneth cells directly sense gut commensals and maintain homeostasis at the intestinal host-microbial interface. Proc Natl Acad Sci USA 2008; 105:20858–20863.

12 Brandl K, Plitas G, Schnabl B, DeMatteo RP, Pamer EG: MyD88-mediated signals induce the bactericidal lectin RegIII gamma and protect mice against intestinal Listeria monocytogenes infection. J Exp Med 2007;204:1891–1900.

13 Kinnebrew MA, Ubeda C, Zenewicz LA, Smith N, Flavell RA, Pamer EG: Bacterial flagellin stimulates Toll-like receptor 5-dependent defense against vancomycin-resistant Enterococcus infection. J Infect Dis 2010;201:534–543.

14 Vaishnava S, Yamamoto M, Severson KM, Ruhn KA, Yu X, Koren O, Ley R, Wakeland EK, Hooper LV: The antibacterial lectin RegIIIgamma promotes the spatial segregation of microbiota and host in the intestine. Science 2011;334:255–258.

15 Brandl K, Plitas G, Mihu CN, Ubeda C, Jia T, Fleisher M, Schnabl B, DeMatteo RP, Pamer EG: Vancomycin-resistant enterococci exploit antibiotic-induced innate immune deficits. Nature 2008;455: 804–807.

16 Round JL, Lee SM, Li J, Tran G, Jabri B, Chatila TA, Mazmanian SK: The Toll-like receptor 2 pathway establishes colonization by a commensal of the human microbiota. Science 2011;332:974–977.

17 Mazmanian SK, Liu CH, Tzianabos AO, Kasper DL: An immunomodulatory molecule of symbiotic bacteria directs maturation of the host immune system. Cell 2005;122:107–118.

18 Mazmanian SK, Round JL, Kasper DL: A microbial symbiosis factor prevents intestinal inflammatory disease. Nature 2008;453:620–625.

19 Hayashi F, Smith KD, Ozinsky A, Hawn TR, Yi EC, Goodlett DR, Eng JK, Akira S, Underhill DM, Aderem A: The innate immune response to bacterial flagellin is mediated by Toll-like receptor 5. Nature 2001;410:1099–1103.

20 Uematsu S, Jang MH, Chevrier N, Guo Z, Kumagai Y, Yamamoto M, Kato H, Sougawa N, Matsui H, Kuwata H, Hemmi H, Coban C, Kawai T, Ishii KJ, Takeuchi O, Miyasaka M, Takeda K, Akira S: Detection of pathogenic intestinal bacteria by Toll-like receptor 5 on intestinal CD11c+ lamina propria cells. Nat Immunol 2006;7:868–874.

21 Feuillet V, Medjane S, Mondor I, Demaria O, Pagni PP, Galan JE, Flavell RA, Alexopoulou L: Involvement of Toll-like receptor 5 in the recognition of flagellated bacteria. Proc Natl Acad Sci USA 2006; 103:12487–12492.

22 Vijay-Kumar M, Sanders CJ, Taylor RT, Kumar A, Aitken JD, Sitaraman SV, Neish AS, Uematsu S, Akira S, Williams IR, Gewirtz AT: Deletion of TLR5 results in spontaneous colitis in mice. J Clin Invest 2007;117:3909–3921.

23 Vijay-Kumar M, Aitken JD, Carvalho FA, Cullender TC, Mwangi S, Srinivasan S, Sitaraman SV, Knight R, Ley RE, Gewirtz AT: Metabolic syndrome and altered gut microbiota in mice lacking Toll-like receptor 5. Science 2010;328:228–231.

24 Letran SE, Lee SJ, Atif SM, Flores-Langarica A, Uematsu S, Akira S, Cunningham AF, McSorley SJ: TLR5-deficient mice lack basal inflammatory and metabolic defects but exhibit impaired CD4 T cell responses to a flagellated pathogen. J Immunol 2011; 186:5406–5412.

25 Carvalho FA, Koren O, Goodrich JK, Johansson ME, Nalbantoglu I, Aitken JD, Su Y, Chassaing B, Walters WA, Gonzalez A, Clemente JC, Cullender TC, Barnich N, Darfeuille-Michaud A, Vijay-Kumar M, Knight R, Ley RE, Gewirtz AT: Transient inability to manage proteobacteria promotes chronic gut inflammation in TLR5-deficient mice. Cell Host Microbe 2012;12:139–152.

26 Ting JP, Lovering RC, Alnemri ES, Bertin J, Boss JM, Davis BK, Flavell RA, Girardin SE, Godzik A, Harton JA, Hoffman HM, Hugot JP, Inohara N, Mackenzie A, Maltais LJ, Nunez G, Ogura Y, Otten LA, Philpott D, Reed JC, Reith W, Schreiber S, Steimle V, Ward PA: The NLR gene family: a standard nomenclature. Immunity 2008;28:285–287.

27 Wilmanski JM, Petnicki-Ocwieja T, Kobayashi KS: NLR proteins: integral members of innate immunity and mediators of inflammatory diseases. J Leukoc Biol 2008;83:13–30.

28 Elinav E, Strowig T, Henao-Mejia J, Flavell RA: Regulation of the antimicrobial response by NLR proteins. Immunity 2011;34:665–679.

29   Martinon F, Burns K, Tschopp J: The inflamma-some: a molecular platform triggering activation of inflammatory caspases and processing of proIL-beta. Mol Cell 2002;10:417–426.

30   Lamkanfi M, Dixit VM: Inflammasomes: guardians of cytosolic sanctity. Immunol Rev 2009;227:95–105.

31   Miao EA, Leaf IA, Treuting PM, Mao DP, Dors M, Sarkar A, Warren SE, Wewers MD, Aderem A: Cas-pase-1-induced pyroptosis is an innate immune ef-fector mechanism against intracellular bacteria. Nat Immunol 2010;11:1136–1142.

32   Tada H, Aiba S, Shibata K, Ohteki T, Takada H: Syn-ergistic effect of Nod1 and Nod2 agonists with Toll-like receptor agonists on human dendritic cells to generate interleukin-12 and T helper type 1 cells. In-fect Immun 2005;73:7967–7976.

33   Ogura Y, Inohara N, Benito A, Chen FF, Yamaoka S, Nunez G: Nod2, a Nod1/Apaf-1 family member that is restricted to monocytes and activates NF-kappaB. J Biol Chem 2001;276:4812–4818.

34   Hisamatsu T, Suzuki M, Reinecker HC, Nadeau WJ, McCormick BA, Podolsky DK: CARD15/NOD2 functions as an antibacterial factor in human intesti-nal epithelial cells. Gastroenterology 2003;124:993–1000.

35   Uehara A, Fujimoto Y, Fukase K, Takada H: Various human epithelial cells express functional Toll-like receptors, NOD1 and NOD2 to produce anti-micro-bial peptides, but not proinflammatory cytokines. Mol Immunol 2007;44:3100–3111.

36   Uehara A, Sugawara Y, Kurata S, Fujimoto Y, Fukase K, Kusumoto S, Satta Y, Sasano T, Sugawara S, Taka-da H: Chemically synthesized pathogen-associated molecular patterns increase the expression of pepti-doglycan recognition proteins via Toll-like recep-tors, NOD1 and NOD2 in human oral epithelial cells. Cell Microbiol 2005;7:675–686.

37   Voss E, Wehkamp J, Wehkamp K, Stange EF, Schro-der JM, Harder J: NOD2/CARD15 mediates induc-tion of the antimicrobial peptide human beta-defen-sin-2. J Biol Chem 2006;281:2005–2011.

38   Gutierrez O, Pipaon C, Inohara N, Fontalba A, Ogu-ra Y, Prosper F, Nunez G, Fernandez-Luna JL: In-duction of Nod2 in myelomonocytic and intestinal epithelial cells via nuclear factor-kappa B activation. J Biol Chem 2002;277:41701–41705.

39   Caetano BC, Biswas A, Lima DS Jr, Benevides L, Mineo TW, Horta CV, Lee KH, Silva JS, Gazzinelli RT, Zamboni DS, Kobayashi KS: Intrinsic expres-sion of Nod2 in CD4+ T lymphocytes is not neces-sary for the development of cell-mediated immunity and host resistance to *Toxoplasma gondii*. Eur J Im-munol 2011;41:3627–3631.

40   Girardin SE, Boneca IG, Viala J, Chamaillard M, La-bigne A, Thomas G, Philpott DJ, Sansonetti PJ: Nod2 is a general sensor of peptidoglycan through mur-amyl dipeptide (MDP) detection. J Biol Chem 2003; 278:8869–8872.

41   Inohara N, Ogura Y, Fontalba A, Gutierrez O, Pons F, Crespo J, Fukase K, Inamura S, Kusumoto S, Hashimoto M, Foster SJ, Moran AP, Fernandez-Lu-na JL, Nunez G: Host recognition of bacterial mur-amyl dipeptide mediated through NOD2. Implica-tions for Crohn's disease. J Biol Chem 2003;278: 5509–5512.

42   Petnicki-Ocwieja T, Hrncir T, Liu YJ, Biswas A, Hudcovic T, Tlaskalova-Hogenova H, Kobayashi KS: Nod2 is required for the regulation of commen-sal microbiota in the intestine. Proc Natl Acad Sci USA 2009;106:15813–15818.

43   Rehman A, Sina C, Gavrilova O, Hasler R, Ott S, Ba-ines JF, Schreiber S, Rosenstiel P: Nod2 is essential for temporal development of intestinal microbial communities. Gut 2011;60:1354–1362.

44   Mondot S, Barreau F, Al Nabhani Z, Dussaillant M, Le Roux K, Dore J, Leclerc M, Hugot JP, Lepage P: Altered gut microbiota composition in immune-im-paired Nod2(–/–) mice. Gut 2011;61:634–635.

45   Cho JH: The genetics and immunopathogenesis of inflammatory bowel disease. Nat Rev Immunol 2008;8:458–466.

46   Wehkamp J, Salzman NH, Porter E, Nuding S, Weichenthal M, Petras RE, Shen B, Schaeffeler E, Schwab M, Linzmeier R, Feathers RW, Chu H, Lima H Jr, Fellermann K, Ganz T, Stange EF, Bevins CL: Reduced Paneth cell alpha-defensins in ileal Crohn's disease. Proc Natl Acad Sci USA 2005;102:18129–18134.

47   Simms LA, Doecke JD, Walsh MD, Huang N, Fowler EV, Radford-Smith GL: Reduced alpha-defensin ex-pression is associated with inflammation and not NOD2 mutation status in ileal Crohn's disease. Gut 2008;57:903–910.

48   Perminow G, Beisner J, Koslowski M, Lyckander LG, Stange E, Vatn MH, Wehkamp J: Defective Paneth cell-mediated host defense in pediatric ileal Crohn's disease. Am J Gastroenterol 2010;105:452–459.

49   Kobayashi KS, Chamaillard M, Ogura Y, Henegariu O, Inohara N, Nunez G, Flavell RA: Nod2-depen-dent regulation of innate and adaptive immunity in the intestinal tract. Science 2005;307:731–734.

50   Biswas A, Liu YJ, Hao L, Mizoguchi A, Salzman NH, Bevins CL, Kobayashi KS: Induction and rescue of Nod2-dependent Th1-driven granulomatous in-flammation of the ileum. Proc Natl Acad Sci USA 2010;107:14739–14744.

51 Biswas A, Petnicki-Ocwieja T, Kobayashi KS: Nod2: a key regulator linking microbiota to intestinal mucosal immunity. J Mol Med (Berl) 2012;90:15–24.

52 Chen GY, Liu M, Wang F, Bertin J, Nunez G: A functional role for Nlrp6 in intestinal inflammation and tumorigenesis. J Immunol 2011;186:7187–7194.

53 Elinav E, Strowig T, Kau AL, Henao-Mejia J, Thaiss CA, Booth CJ, Peaper DR, Bertin J, Eisenbarth SC, Gordon JI, Flavell RA: NLRP6 inflammasome regulates colonic microbial ecology and risk for colitis. Cell 2011;145:745–757.

54 Normand S, Delanoye-Crespin A, Bressenot A, Huot L, Grandjean T, Peyrin-Biroulet L, Lemoine Y, Hot D, Chamaillard M: Nod-like receptor pyrin domain-containing protein 6 (NLRP6) controls epithelial self-renewal and colorectal carcinogenesis upon injury. Proc Natl Acad Sci USA 2011;108:9601–9606.

55 Henao-Mejia J, Elinav E, Jin C, Hao L, Mehal WZ, Strowig T, Thaiss CA, Kau AL, Eisenbarth SC, Jurczak MJ, Camporez JP, Shulman GI, Gordon JI, Hoffman HM, Flavell RA: Inflammasome-mediated dysbiosis regulates progression of NAFLD and obesity. Nature 2012;482:179–185.

Koichi S. Kobayashi MD, PhD
Department of Microbial and Molecular Pathogenesis, College of Medicine
Texas A&M Health Science Center, 415A Reynolds Medical Building
College Station, TX 77843-1114 (USA)
E-Mail kobayashi@medicine.tamhsc.edu

Guarino A, Quigley EMM, Walker WA (eds): Probiotic Bacteria and Their Effect on Human Health and Well-Being.
World Rev Nutr Diet. Basel, Karger, 2013, vol 107, pp 43–55 (DOI: 10.1159/000345733)

Importance of Early Microbial Colonization for Intestinal Immune Development

P. Brandtzaeg[a, b]

[a]Laboratory for Immunohistochemistry and Immunopathology (LIIPAT), Centre for Immune Regulation (CIR), University of Oslo, and [b]Department of Pathology, Oslo University Hospital, Rikshospitalet, Oslo, Norway

Abstract

Many variables influence the development of secretory immunity and oral tolerance – two immuno-logical mechanisms that are of paramount importance for the intestinal barrier function and immune homeostasis. Increased epithelial permeability is likely a significant primary or secondary event in the pathogenesis of several intestinal disorders, including adverse immune reactions to food proteins and commensal bacteria. This barrier variable is determined by the developmental stage (e.g. preterm vs. term newborn, infant vs. adult), concurrent infections, and the shielding effect of secretory IgA (SIgA) antibodies provided by breast milk or the individual's intestinal immune system. The clinical consequences will depend on how fast 'closure' of the epithelial barrier can be attained or reestablished, which is influenced both by the individual's age and by successful mounting of adaptive SIgA responses as well as generation of oral tolerance (mucosally induced hyporesponsiveness) against innocuous antigens from the diet and components of the normal indigenous microbiota. Generation of SIgA is the best defined effector mechanism of the intestinal immune system; its enhancement and homeostatic immune regulation induced by commensal bacteria is therefore of considerable clinical interest. Also, the intestinal induction of regulatory T cells (that are of central importance in oral tolerance) depends on an adequate development of a complex gut microbiota. Importantly, the feeding and treatment regimen (e.g. antibiotics) to which the infant is subjected, and apparently also the mode of delivery, may perturb the balance of the gut bacteria and thereby jeopardize the homeostasis of the developing mucosal immune system.

The epithelial lining of the gut includes numerous folds, crypts (glands), and villi with brush border – together covering an area 100–150 times larger than the skin. This monolayered epithelium constitutes a vulnerable physical barrier between the exogenous environment and the body's internal milieu; it is therefore protected by numerous innate defenses intimately cooperating with adaptive mucosal immunity which largely relies on secretory immunoglobulin A (SIgA) and SIgM antibodies. The former antibody class provides the most extensive anti-inflammatory defense of the body by performing immune exclusion – a term coined for mucosal control of microbial colonization and restriction of antigen uptake [1]. SIgA also maintains host mutualism with the indigenous microbiota [2].

Components of the gut microbiota activate not only the secretory antibody system, but also suppressive mechanisms, mainly depending on regulatory T (T_{Reg}) cells [3]. Such mucosally induced hyporesponsiveness induced via the gut is called 'oral tolerance' and probably explains why most individuals show no adverse immune reactions against harmless dietary and environmental proteins [4]. Oral tolerance depends on a well-functioning epithelial barrier, maintained by exposure to immune-modulating microbial components with conserved microbe-associated molecular patterns (MAMPs). The MAMPs interact with germline-encoded pattern recognition receptors expressed by the epithelium and enhance oral tolerance via tolerogenic antigen-presenting dendritic cells (DCs) [3, 4].

The mucosae are favored as the portal of entry by pathogens and allergens, thus epithelial defense is crucial to health. The neonatal period is particularly critical as adaptive mucosal immunity is virtually lacking during a variable period after birth. Breastfeeding is therefore important by providing secretory antibodies and immune-modulating factors to reinforce the baby's developing immune system.

Birth marks the transition from an essentially sterile fetal environment to one that is rich in bacteria and nutritional or other exogenous substrates [5]. This massive microbial encounter seems to determine the postnatal intestinal colonization: after natural birth, the bacterial composition resembles that of the maternal vaginal and gut microbiota, whereas after cesarean section, the infant's intestinal microbiota includes a large number of bacteria from the external environment. With weaning, an increasingly diverse microbiota is established that is highly individual and remains relatively stable throughout life [6]. This review will focus on how the early microbial colonization influences adaptive and innate immune mechanisms in the gut, including the epithelial barrier function.

Postnatal Maturation of the Adaptive Intestinal Immune System

Secretory antibodies originate from B cell responses elicited primarily in organized mucosa-associated lymphoid tissue, which samples antigens directly from epithelial surfaces [7]. After activation, the immune cells are disseminated to mucosal effector sites such as the gut lamina propria – a 'homing' process depending on adhesion molecules and chemokines/chemokine receptors guiding site-specific cellular extravasation [7].

Gut-associated lymphoid tissue (GALT) is comprised chiefly of aggregated (Peyer's patches) and isolated lymphoid follicles (ILFs). Most human Peyer's patches are located in the distal ileum, whereas ILFs occur mainly in the large bowel [7]. All components of GALT, including ILFs and the appendix, are believed to be functionally similar; they contain a characteristic follicle-associated epithelium (FAE) with specialized 'micro-fold' or 'membrane' (M) cells capable of transporting live and dead antigens from the lumen into the lymphoid tissue – a process aided by DCs scattered throughout FAE [7]. The development of M cells depends on exposure of FAE to bacteria, as shown in germ-free rats after reconstitution of a conventional pathogen-free microbiota [8].

The lymphoid structures of Peyer's patches and the draining mesenteric lymph nodes are formed before birth with discrete T and B cell areas being apparent after 19 weeks' gestation in humans [9]; however, the size of these structures and germinal centers of B cell follicles depend on the postnatal microbial colonization. Also, ILFs and cryptopatches (the latter seen only in mice) cannot be observed until after birth (fig. 1a). B cells of GALT germinal centers express mainly surface IgA as a result of Ig heavy-chain gene switching in the course of B cell differentiation to IgA-producing plasmablasts. Notably, IgA induction is much more prominent in GALT than in other mucosa-associated lymphoid tissue structures [7].

In parallel with the bacterial colonization, the homing of lymphocytes (including IgA+ plasmablasts) to the gut lamina propria seems to follow defined kinetics – apparently reflecting a series of endogenous and exogenous signals regulating intestinal postnatal immune maturation. Innate cells – such as lymphoid tissue inducer cells, natural killer (NK) cells, NK-like NKp46+ cells, and T helper 2 (Th2)-like cells – migrate during the first 4 weeks after birth from the murine fetal liver to the gut mucosa driven by endogenous signals [5]. By contrast, enhanced recruitment of CD8+ intraepithelial lymphocytes (IELs) and forkhead box protein 3 (FOXP3)+ T_{Reg} cells to the rodent gut as well as the production of anti-inflammatory IL-10 have been associated with bacterial colonization [5, 10]. The T_{Reg} cells help to keep proinflammatory CD4+ Th cells and cytotoxic CD8+ T cells under control to preserve the epithelial barrier [3].

The delay of the postnatal mucosal immune activation parallels a temporary immaturity of systemic immunity [11]. Very few plasmablasts occur in peripheral blood of newborns [12], but after 1 month those with IgA-producing capacity (presumably GALT-derived) are remarkably increased [13], signifying progressive microbial stimulation (fig. 1a; table 1). An early elevation of plasmablasts can be seen in preterm infants, especially in those with intrauterine infections, although IgM production dominates in such cases [12].

In agreement with these observations, only scattered IgM+ (and IgG+) intestinal plasma cells could be seen in newborns, and IgA+ cells were either absent or extremely rare even at 10 days of age [11]. The numbers of IgM- and IgA-producing cells increased rapidly after 2–4 weeks – the latter becoming predominant at 1–2 months, usually peaking around 12 months (fig. 1b). However, in affluent societies it may take several years for the size of the IgA+ plasma cell population to reach that of healthy adults, whereas a fast postnatal increase of SIgA were observed in children living in developing countries with a heavy microbial load [3]. Similarly, the number of intes-

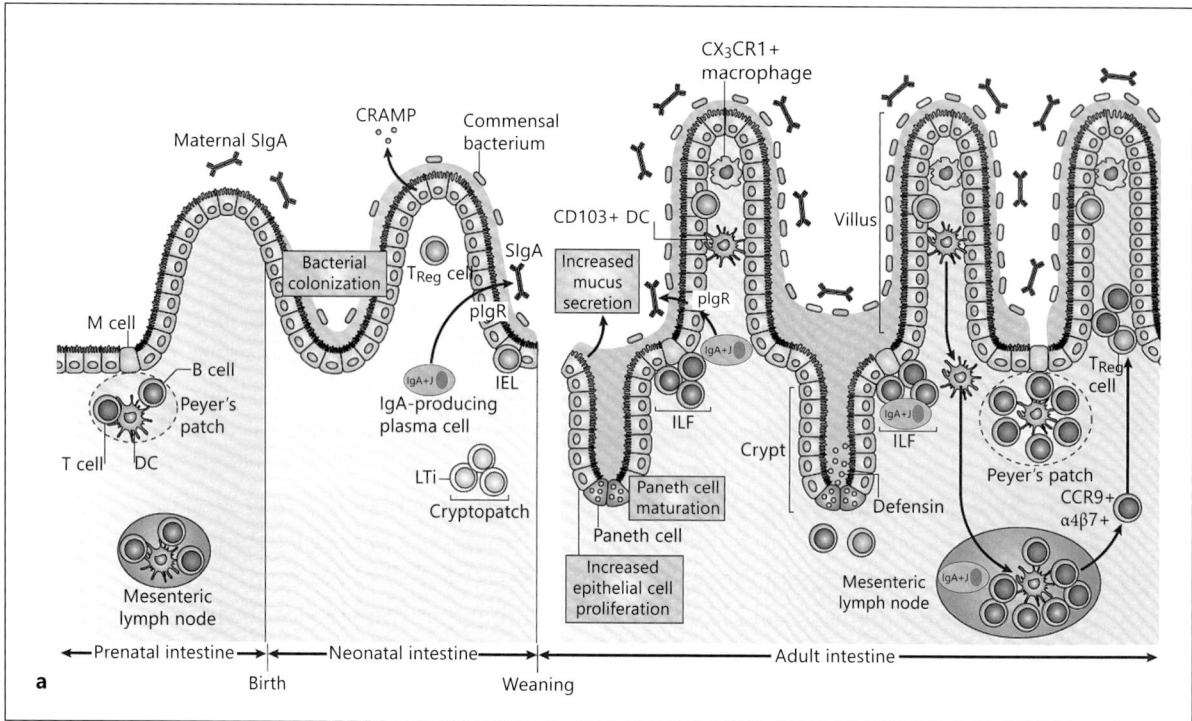

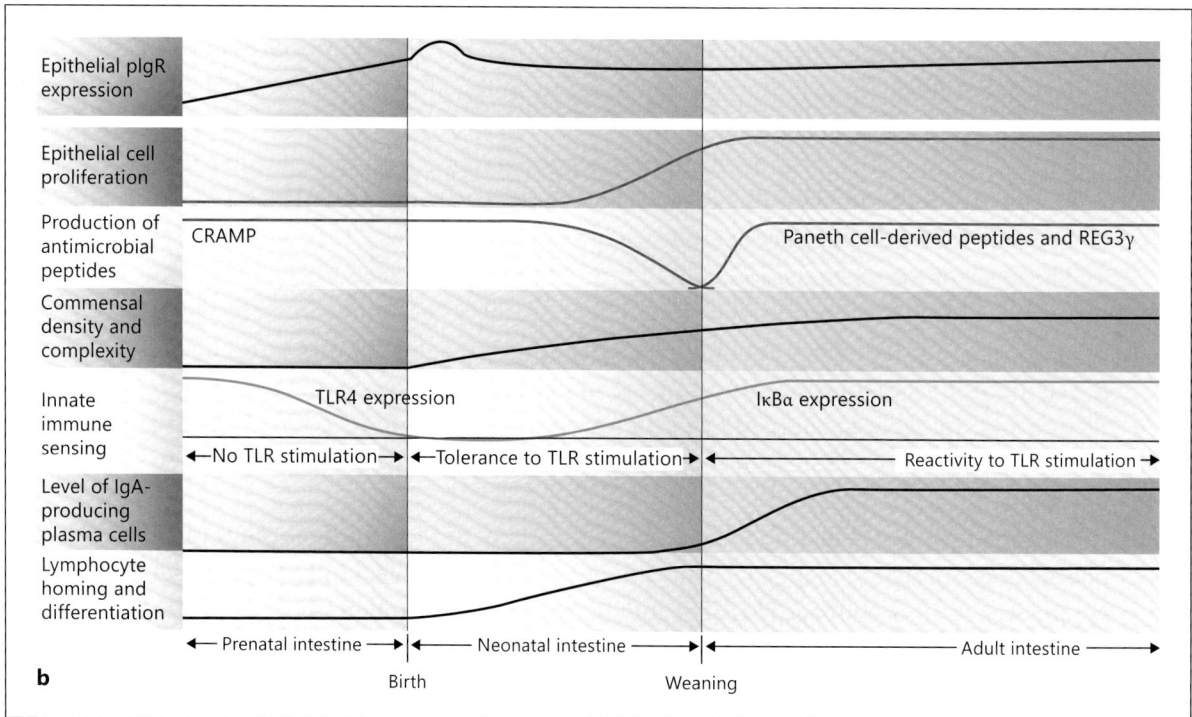

Fig. 1.

tinal IgA+ cells normalized 4 weeks after colonization of germ-free mice with commensal bacteria [14, 15], and dependency on a complex gut microbiota was documented (table 2). A recent study showed that an optimal stimulatory effect on mucosal IgA production required a host-specific microbiota, and the same held true for small-intestinal T cell activity [16].

Interestingly, pioneering studies in mice showed that the microbiota stimulates a self-limiting intestinal SIgA response [17]. Such transient SIgA production is probably necessary to allow access of microbial constituents to GALT. In this manner, it seems that the intestinal IgA response is continuously adapting to the changing microbiota [18], which would be especially relevant in the early postnatal period.

Developmental Modulation of Epithelial Antibody Translocation

The importance of GALT for adaptive immunity is revealed by the fact that at least 80% of all plasmablasts and plasma cells of the body are located in the intestinal mucosa, amounting to approximately 10^{10} cells/m of normal adult gut [1]. Some 90% of these terminally differentiated memory/effector B cells produce polymers of IgA (dimers and some trimers, collectively called pIgA), which are efficiently transported externally as

Fig. 1. a, **b** Development and maturation of the intestinal immune system and the epithelial barrier. Inherent and environmental signals drive the mucosal changes observed both in mice and humans; the postnatal establishment of an increasingly complex and dense gut microbiota is a decisive variable. **a** The lymphoid structures of Peyer's patches and mesenteric lymph nodes are generated before birth, but mature during the postnatal period. By contrast, cryptopatches (seen only in mice) and ILFs are formed after birth. Specialized antigen-sampling epithelial cells, known as M cells, reside above Peyer's patches and ILFs and facilitate antigen transport from the gut lumen to the underlying lymphoid cells. Simultaneously, innate lymphocytes, such as lymphoid tissue inducer (LTi) cells, as well as adaptive T cells leave the liver and thymus, respectively, and colonize the mucosa, including the epithelium. IELs reside in close proximity to epithelial cells. Also, increasing numbers of CD103+ DCs and CX$_3$CR1+ macrophage-like cells home to the gut mucosa. In contrast to innate lymphocytes, T$_{Reg}$ cells populate the intestinal mucosa in response to bacterial colonization. Although B cells are present in gut tissue during early development, plasma cells producing dimeric IgA are only generated after birth to provide SIgA which is transported to the lumen by the pIgR. Maternal SIgA is provided by breast milk during the early postnatal period. **b** The neonatal mucosa is characterized by little epithelial cell proliferation, absence of developed crypts (intestinal glands) and crypt-based Paneth cells, but marked expression of CRAMP; by contrast, the formation of intestinal crypts late during the second week after birth in mice initiates increased proliferation and rapid epithelial cell renewal, generation of α-defensin-producing Paneth cells, and upregulation of the antibacterial C-type lectin REG3γ. A decrease in the epithelial expression of TLR4 before birth and a steady increase in the level of the NF-κB inhibitor IκBα during the postnatal period reduce the responsiveness to bacterial lipopolysaccharide and other proinflammatory stimuli. Such acquisition of epithelial TLR tolerance creates a neonatal period of decreased innate immune responsiveness. Note that the small intestinal epithelium at birth has a more mature phenotype in humans than in mice. Modified from Renz et al. [5].

Table 1. Increase of Ig-secreting cells (Ig-SCs) detected by ELISPOT in peripheral blood of healthy neonates during the first month of life

Day of life	IgA-SCs/10^6 PBMCs	Increment (%) of positive samples		
		IgA	IgM	IgG
0–5	<8*			
6–14		58	38	46
15–21		67	33	40
22–31	~500**	78	31	39

* Data from Stoll et al. [12]; ** Data from Nahmias et al. [13].

Table 2. Effect of fecal bacteria from growing conventional (CV) mice on the development of intestinal IgA+ plasma cells in adult germ-free (GF) mice

GF mice colonized with fecal microflora from:	IgA+ plasma cell (number/villus)[1]
Adult CV mice	41±1
Adult GF mice	4±0.5
Growing CV mice	
1–4 days old	15±2
7–23 days old	23±1
25 days old (4 days after weaning)	43±1

[1] Mean number ± SEM 4 weeks after bacterial colonization. Data from Moreau et al. [15].

SIgA antibodies by an epithelial glycoprotein of approximately 100 kDa called membrane secretory component (SC), now generally referred to as the polymeric Ig receptor (pIgR). This transport mechanism (fig. 2a) is shared by pIgA and pentameric IgM, which both contain a 15-kDa polypeptide called the joining (J) chain – abundantly produced by the mucosal plasma cells [19]. The J chain not only covalently links together the subunits of the Ig polymers, but it also constitutes part of the pIgR-binding site [1, 20].

After transcytosis of the Ig polymers to the luminal epithelial face, SIgA and SIgM are extruded by cleavage of pIgR (fig. 2a) – only its C-terminal small domain remains for apical degradation while the 80-kDa extracellular part is incorporated into the secretory antibodies as bound SC and confers protection against degradation [21], particularly of SIgA where it becomes covalently linked [1]. Milk SIgA antibodies can therefore to some extent survive passage through the infant's gut.

The epithelial production of pIgR/SC begins in fetal life around 3–5 months and is constitutively regulated [11]; its expression increases steadily until birth and may be followed by a postnatal peak (fig. 1b), which could reflect microbial encounter. Thus, studies with the polarized human colon carcinoma cell line HT-29 have shown that cytokines from antigen-activated Th cells and antigen-presenting cells upregulate

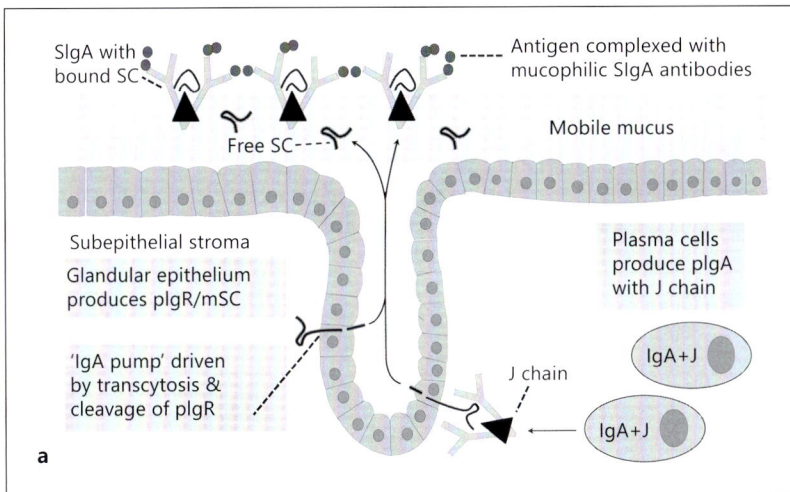

Fig. 2. a,b Receptor-mediated epithelial export of polymeric IgA (pIgA, mainly dimers) to provide SIgA antibodies and microbiota-induced enhancement of the gut epithelial barrier function. **a** At the mucosal surface, SIgA antibodies together with mucus perform immune exclusion of antigens. The epithelial pIgR is expressed basolaterally, mainly in the intestinal crypts (glands), as membrane secretory component (mSC) and mediates external transcytosis of pIgA (and pentameric IgM, not shown). SIgA is released to the lumen with bound SC by apical cleavage of pIgR, in the same manner as unoccupied pIgR (carrying no ligand) is cleaved to provide free SC. Mucosal plasma cells produce abundantly pIgA with incorporated J chain (IgA+J), which is required for high-affinity epithelial binding of the pIgR ligands.

pIgR/SC expression, and the export of secretory antibodies in response to microbial stimulation is thereby enhanced [3]. A positive effect has also been shown for butyrate, particularly in combination with various cytokines (fig. 2b). This short-chain fatty acid is an anaerobic microbial fermentation product of oligosaccharides and an important energy source of colonic epithelial cells. Moreover, interactions at the apical epithelial side of pattern recognition receptors such as the Toll-like receptors TLR3 and TLR4 with their ligands (double-stranded RNA and lipopolysaccharide) upregulate pIgR/SC expression [3].

In adult humans, on average approximately 40 mg/kg body weight of pIgA is exported to the gut lumen every day, which is more than the total daily production of IgG in the body [1]. As a consequence, the epithelial pIg transport is often referred to as the 'IgA pump'. Excess of unoccupied pIgR, on average about 50% of the amount produced, is cleaved and released to the lumen by the same mechanism to form so-called free SC (fig. 2a). This 80-kDa fragment occurs in most secretions, including breast milk. In IgA deficiency and in the newborn period, it is important that free SC, by equilibrium with bound SC, exerts a stabilizing effect on the quaternary structure of SIgM, in which SC remains noncovalently linked [1]. In addition, both free and bound SC exhibit several innate defense functions [21].

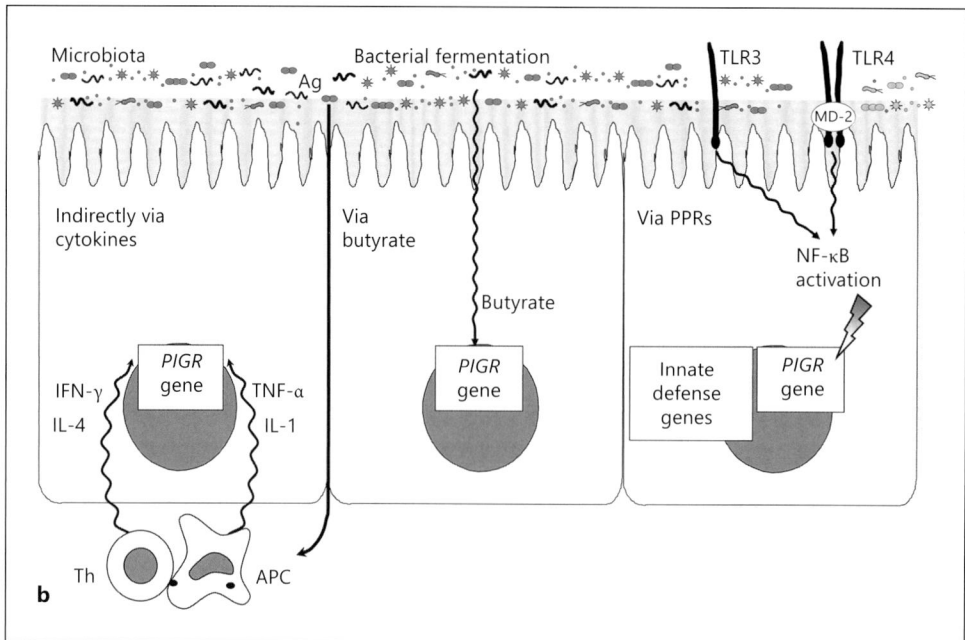

Fig. 2. b Schematic illustrations of three possible ways in which the expression of the pIgR gene locus *PIGR* and innate defense genes may be upregulated. Bacterial components may provide signals through Toll-like receptors such as TLR3 and TLR4, leading to epithelial NF-κB activation. Other pattern recognition receptors may also be involved and enhance the expression of a variety of epithelial defense genes. The pIgR-enhancing effect of the cytokines IFN-γ, IL-4, TNF-α, and IL-1 provided by antigen (Ag)-activated antigen-presenting cells (APCs) and Th cells, can be enhanced by the vitamin A derivative retinoic acid (not shown). Details are discussed in the text. MD-2 = Molecule associated with TLR4 and responsible for its lipopolysaccharide signaling; PPRs = Pattern recognition receptors. Adapted from Brandtzaeg [3].

Ontogeny of Epithelial Innate Immune Properties

The neonatal gut varies among species in terms of maturity, depending in part on the length of the gestation period. The small intestinal mucosa of newborn humans has a mature crypt-villous architecture, with continuous stem cell proliferation, and epithelial cell migration and differentiation. In mice, small intestinal crypts only develop 10–12 days after birth, accompanied by increased epithelial cell renewal and transcriptional reprogramming of enterocytes, which includes changed expression of genes involved in nutrient transport, metabolism, and cell differentiation [5]. Also, the enteric spectrum of antimicrobial peptides changes significantly during neonatal development in mice (fig. 1b). During the first 2 weeks of postnatal life, when mature crypt-based Paneth cells are absent, the mouse intestinal epithelium expresses cathelicidin-related antimicrobial peptide, or CRAMP. The Paneth cells start to produce defensins at weaning, and then the CRAMP expression decreases.

While defensin production by Paneth cells is independent on bacterial colonization, expression of other antimicrobial peptides by epithelial cells – such as the C-type lectin regenerating islet-derived protein 3γ (REG3γ) – requires a microbiota (fig. 1b) and is supported by IL-22-producing RORγt+NKp46+ lymphocytes [5]. Thus, administration to mice of broad-spectrum antibiotics by gavage has been shown to reduce significantly the colonic epithelial expression of 70 genes; and 5 of the 7 genes that were more than fourfold less active than normal encoded antimicrobial peptides including Ang4, Pla-2g2a, Retnlb, REG3γ, and REG3β [3]. Furthermore, REG3γ has been detected in murine γδ IELs in response to microbial stimulation of epithelial cell-intrinsic adapter protein MyD88 signaling, probably mediated through Toll-like receptors [22, 23]. Human γδ IELs also seem to belong to the first-line innate mucosal defense as deemed from their behavior in AIDS patients and lack of response to antiretroviral treatment [24].

The severe consequences of manipulating mouse microbiota by antibiotic treatment have recently been reported in terms of susceptibility to virus infection [25]. Importantly, changes in the composition of antibacterial peptides and their activity have also been observed in the neonatal human gut [26]. As enteric antimicrobial peptide synthesis is associated with the composition of the microbiota and chronic mucosal inflammation, these processes during the postnatal period might significantly affect susceptibility to inflammatory disease later in life [27].

Neonatal Crosstalk between Microbiota and Epithelium

Exposure of mouse pups during or shortly after natural birth to endotoxin/lipopolysaccharide from Gram-negative bacteria induces a transient transcriptional activation of the small intestinal epithelium with upregulated expression of microRNA-146a. Increased levels of microRNA-146a in intestinal epithelial cells cause translational repression of the Toll-like receptor signaling molecule IL-1-associated kinase 1 (IRAK1). This, together with the proteasomal degradation of IRAK1, contributes to innate immune tolerance by inhibiting Toll-like receptor signaling [5]. Other mechanisms that help to prevent inappropriate immune stimulation by Toll-like receptor agonists during the postnatal intestinal colonization include downregulation of TLR4 in the intestinal epithelium – the expression of which is high in late fetal life of mice but decreases at birth [28]. Conversely, the epithelial expression of the NF-κB inhibitor IκBα steadily increases during the postnatal period [29]. The combination of decreasing levels of TLR4 and increasing levels of IκBα in the epithelial cells effectively enhances the threshold of immune activation in the gut epithelium (fig. 1b).

Notably, decreased IRAK1 protein expression in mouse neonatal epithelium requires continuous Toll-like receptor signaling. This facilitates prolonged upregulation of microRNA-146a expression and simultaneously induces sustained expression of genes supporting cell maturation, survival, and nutrient absorption [30]. Innate immune signaling by epithelial cells seems to be essential for immune tolerance, as lack of the proinflammatory signaling molecule transforming growth factor-β (TGF-β)-activated kinase 1 (TAK1) specifically in the murine intestinal epithelium leads to early inflammation, tissue damage, and postnatal mortality [31]. Thus, although inappropriate stimulation of the neonatal innate immune system by the microbiota must be prevented, controlled innate immune activation significantly contributes to nutrient absorption, angiogenesis, epithelial cell differentiation, and barrier reinforcement [5].

Despite the decreased sensitivity of epithelial cells to Toll-like receptor stimulation in murine neonates, other innate immune signaling pathways remain fully functional. For example, rotavirus infection of the intestinal epithelium in neonatal mice is efficiently sensed by the helicases retinoic acid-inducible gene I (RIG-I) and melanoma differentiation-associated gene 5 (MDA5) [5]. Confronted with the colonization of the gut microbiota, the neonatal intestinal epithelium seems to calibrate its bacterial sensitivity and modify signaling pathways after initial stimulation while maintaining antiviral host defenses. The mucus layer plays an important role in the microbiota-epithelial cross-talk by keeping the bacteria away from the epithelial cells (fig. 2), but little is known about its function in the early postnatal period [3]. The homeostatic function of SIgA is enhanced by its mucophilic properties and reflected by the fact that it coats a substantial proportion of commensal bacteria [3], even in young children [32].

Peripheral Induction of Regulatory T Cells

There may be certain 'windows of opportunity' for the induction of oral tolerance, particularly during the perinatal period. Thus, peripheral generation of T_{Reg} cells seems to be part of the immune maturation process, depending mainly on a finely tuned cross-talk between innate and adaptive immunity, as well as on the epithelial integrity [3]. Interestingly, T_{Reg} cells are abundant in the human gut at birth [33] and in fetal mesenteric lymph nodes [34], and their homing to the gut mucosa seems to be particularly active in infancy [35]. The increasing prevalence of allergic and inflammatory disorders in infancy suggests that immune dysregulation leading to hypersensitivity is mainly an early event that compromises the function of immature antigen-presenting cells and T_{Reg} cells [4]. This could be associated with a perturbed microbiota (dysbiosis) which to some extent might be ascribed to delivery by cesarian section or use of antibiotics [6]. Also, reduced commensal SIgA coating together with dysbiosis has been observed in children with celiac disease [32].

Peripheral T_{Reg} cell generation apparently depends on appropriate stimulation of mucosal DCs by certain MAMPs derived from commensal bacteria, which induce the signaling molecules and transcription factors dictating the differentiation pathways and cytokine profiles of the activated T cells [4, 36]. Mechanistically, it is thought that CD103+CD7+ migratory DCs continuously carry antigens/MAMPs from the gut to the mesenteric lymph nodes where they promote the induction of T_{Reg} cells in the presence of IL-2, TGF-β, IL-10, and vitamin A-derived retinoic acid. In addition, subepithelial nonmigratory CD103–CX$_3$CR1+ macrophage-like cells produce IL-10, which supports proliferation of the T$_{Reg}$ cells when they have homed to the lamina propria to exert their suppressive function [3].

Murine colonic T$_{Reg}$ cells are largely specific for antigens derived from commensals, suggesting that the T$_{Reg}$ cell repertoire is shaped significantly by the local antigenic environment in a process of peripheral immune education [37]. Such extrathymic generation of T$_{Reg}$ cells has also been observed in the gut of mice fed a soluble protein antigen. Additionally, there is apparently a reciprocal effect on preserving the composition of the gut microbiota, probably because the local induction of T$_{Reg}$ cells maintains a healthy mucosa [38].

A prominent member of the gut microbiota in human infants, *Bifidobacterium infantis,* was shown to markedly induce FOXP3+ T$_{Reg}$ cells after deliberate consumption in mice [39]. Notably, neonatal CD4+ T cells in mice are prone to differentiate into T$_{Reg}$ cells following stimulation [40], as are human cord blood cells, probably as a result of perinatal exposure to maternal progesterone [41]. Later in development, members of the *Clostridium* cluster IV and XIVa might take over the role of *B. infantis* in promoting the local expansion of T$_{Reg}$ cells in the colon [23]. The induced anti-inflammatory response might partially depend on release of TGF-β from IELs [42]. *Bacteroides fragilis* also seems to have unique T$_{Reg}$ cell-inducing and epithelium-associating properties [23].

Conclusions

SIgA is the most prominent effector component of the intestinal immune system; its adaptive enhancement and the mucosal immune regulation involving T$_{Reg}$ cells – both induced by commensal bacteria in the newborn period – are of considerable clinical interest. Importantly, the feeding and treatment regimen (e.g. antibiotics) to which the infant is subjected, may disturb the balance of the developing gut microbiota [6] and thereby the homeostasis of the mucosal immune system.

Acknowledgements

The author is grateful to Hege Eliassen for excellent secretarial assistance. Studies at LIIPAT were supported by the Research Council of Norway, the University of Oslo, and Oslo University Hospital.

References

1 Brandtzaeg P: Mucosal immunity: induction, dissemination, and effector functions. Scand J Immunol 2009;70:505–515.

2 Peterson DA, McNulty NP, Guruge JL, Gordon JI: IgA response to symbiotic bacteria as a mediator of gut homeostasis. Cell Host Microbe 2007;2:328–339.

3 Brandtzaeg P: Gate-keeper function of the intestinal epithelium. Benef Microbes 2013;4:67–82.

4 Brandtzaeg P: Food allergy: separating the science from the mythology. Nat Rev Gastroenterol Hepatol 2010;7:380–400.

5 Renz H, Brandtzaeg P, Hornef M: The impact of perinatal immune development on mucosal homeostasis and chronic inflammation. Nat Rev Immunol 2012;12:9–23.

6 Cho I, Blaser MJ: The human microbiome: at the interface of health and disease. Nat Rev Genet 2012;13:260–270.

7 Brandtzaeg P: Functions of mucosa-associated lymphoid tissue in antibody formation. Immunol Invest 2010;39:303–355.

8 Yamanaka T, Helgeland L, Farstad IN, Fukushima H, Midtvedt T, Brandtzaeg P: Microbial colonization drives lymphocyte accumulation and differentiation in the follicle associated epithelium of Peyer's patches. J Immunol 2003;170:816–822.

9 Spencer J, MacDonald TT: Development and function of human intraepithelial lymphocytes; in Kiyono H and McGhee JR (eds): Mucosal Immunology: Intraepithelial Lymphocytes, Advances in Host Defense Mechanisms. New York, Raven Press, 1994, vol 9, pp 136–146.

10 Brandtzaeg P, Farstad IN, Helgeland L: Phenotypes of T cells in the gut. Chem Immunol 1998;71:1–26.

11 Brandtzaeg P, Nilssen DE, Rognum TO, Thrane PS: Ontogeny of the mucosal immune system and IgA deficiency. Gastroenterol Clin North Am 1991;20:397–439.

12 Stoll BJ, Lee FK, Hale E, Schwartz D, Holmes R, Ashby R, Czerkinsky C, Nahmias AJ: Immunoglobulin secretion by the normal and the infected newborn infant. J Pediatr 1993;122:780–786.

13 Nahmias A, Stoll B, Hale E, Ibegbu C, Keyserling H, Innis-Whitehouse W, Holmes R, Spira T, Czerkinsky C, Lee F: IgA-secreting cells in the blood of premature and term infants: normal development and effect of intrauterine infections. Adv Exp Med Biol 1991;310:59–69.

14 Crabbé PA, Nash DR, Bazin H, Eyssen H, Heremans JF: Immunohistochemical observations on lymphoid tissues from conventional and germ-free mice. Lab Invest 1970;22:448–457.

15 Moreau MC, Raibaud P, Muller MC: Relationship between the development of the intestinal IgA immune system and the establishment of microbial flora in the digestive tract of young holoxenic mice (in French). Ann Immunol (Paris) 1982;133D: 29–39.

16 Chung H, Pamp SJ, Hill JA, Surana NK, Edelman SM, Troy EB, Reading NC, Villablanca EJ, Wang S, Mora JR, Umesaki Y, Mathis D, Benoist C, Relman DA, Kasper DL: Gut immune maturation depends on colonization with a host-specific microbiota. Cell 2012;149:1578–1593.

17 Shroff KE, Meslin K, Cebra JJ: Commensal enteric bacteria engender a self-limiting humoral mucosal immune response while permanently colonizing the gut. Infect Immun 1995;63:3904–3913.

18 Hapfelmeier S, Lawson MA, Slack E, Kirundi JK, Stoel M, Heikenwalder M, Cahenzli J, Velykoredko Y, Balmer ML, Endt K, Geuking MB, Curtiss R 3rd, McCoy KD, Macpherson AJ: Reversible microbial colonization of germ-free mice reveals the dynamics of IgA immune responses. Science 2010;328:1705–1709.

19 Brandtzaeg P: Presence of J chain in human immunocytes containing various immunoglobulin classes. Nature 1974;252:418–420.

20 Brandtzaeg P, Prydz H: Direct evidence for an integrated function of J chain and secretory component in epithelial transport of immunoglobulin. Nature 1984;311:71–73.

21 Corthésy B: Role of secretory immunoglobulin A and secretory component in the protection of mucosal surfaces. Future Microbiol 2010;5:817–829.

22 Ismail AS, Severson KM, Vaishnava S, Behrendt CL, Yu X, Benjamin JL, Ruhn KA, Hou B, DeFranco AL, Yarovinsky F, Hooper LV: γδ Intraepithelial lymphocytes are essential mediators of host-microbial homeostasis at the intestinal mucosal surface. Proc Natl Acad Sci USA 2011;108:8743–8748.

23 Smith PM, Garrett WS: The gut microbiota and mucosal T cells. Front Microbiol 2011;2:111.

24 Nilssen DE, Brandtzaeg P: Intraepithelial γδ T cells remain increased in the duodenum of AIDS patients despite antiretroviral treatment. PLoS One 2012; 7:e29066.

25 Abt MC, Osborne LC, Monticelli LA, Doering TA, Alenghat T, Sonnenberg GF, Paley MA, Antenus M, Williams KL, Erikson J, Wherry EJ, Artis D: Commensal bacteria calibrate the activation threshold of innate antiviral immunity. Immunity 2012;37:158–170.

26 Kai-Larsen Y, Bergsson G, Gudmundsson GH, Printz G, Jörnvall H, Marchini G, Agerberth B: Antimicrobial components of the neonatal gut affected upon colonization. Pediatr Res 2007;61:530–536.

27 Salzman NH, Hung K, Haribhai D, Chu H, Karlsson-Sjöberg J, Amir E, Teggatz P, Barman M, Hayward M, Eastwood D, Stoel M, Zhou Y, Sodergren E, Weinstock GM, Bevins CL, Williams CB, Bos NA: Enteric defensins are essential regulators of intestinal microbial ecology. Nat Immunol 2010;11: 76–83.

28 Gribar SC, Sodhi CP, Richardson WM, Anand RJ, Gittes GK, Branca MF, Jakub A, Shi XH, Shah S, Ozolek JA, Hackam DJ: Reciprocal expression and signaling of TLR4 and TLR9 in the pathogenesis and treatment of necrotizing enterocolitis. J Immunol 2009;182:636–646.

29 Claud EC, Lu L, Anton PM, Savidge T, Walker WA, Cherayil BJ: Developmentally regulated IκB expression in intestinal epithelium and susceptibility to flagellin-induced inflammation. Proc Natl Acad Sci USA 2004;101:7404–7408.

30 Chassin C, Kocur M, Pott J, Duerr CU, Gütle D, Lotz M, Hornef MW: miR-146a mediates protective innate immune tolerance in the neonate intestine. Cell Host Microbe 2010;8:358–368.

31 Kajino-Sakamoto R, Inagaki M, Lippert E, Akira S, Robine S, Matsumoto K, Jobin C, Ninomiya-Tsuji J: Enterocyte-derived TAK1 signaling prevents epithelium apoptosis and the development of ileitis and colitis. J Immunol 2008;181:1143–1152.

32 De Palma G, Nadal I, Medina M, Donat E, Ribes-Koninckx C, Calabuig M, Sanz Y: Intestinal dysbiosis and reduced immunoglobulin-coated bacteria associated with coeliac disease in children. BMC Microbiol 2010;10:63.

33 Weitkamp JH, Rudzinski E, Koyama T, Correa H, Matta P, Alberty B, Polk DB: Ontogeny of FOXP3+ regulatory T cells in the postnatal human small intestinal and large intestinal lamina propria. Pediatr Dev Pathol 2009;12:443–449.

34 Michaëlsson J, Mold JE, McCune JM, Nixon DF: Regulation of T cell responses in the developing human fetus. J Immunol 2006;176:5741–5748.

35 Grindebacke H, Stenstad H, Quiding-Järbrink M, Waldenström J, Adlerberth I, Wold AE, Rudin A: Dynamic development of homing receptor expression and memory cell differentiation of infant CD4⁺CD25high regulatory T cells. J Immunol 2009; 183:4360–4370.

36 Hooper LV, Macpherson AJ: Immune adaptations that maintain homeostasis with the intestinal microbiota. Nat Rev Immunol 2010;10:159–169.

37 Lathrop SK, Bloom SM, Rao SM, Nutsch K, Lio CW, Santacruz N, Peterson DA, Stappenbeck TS, Hsieh CS: Peripheral education of the immune system by colonic commensal microbiota. Nature 2011;478: 250–254.

38 Josefowicz SZ, Niec RE, Kim HY, Treuting P, Chinen T, Zheng Y, Umetsu DT, Rudensky AY: Extrathymically generated regulatory T cells control mucosal T_H2 inflammation. Nature 2012;482:395–399.

39 O'Mahony C, Scully P, O'Mahony D, Murphy S, O'Brien F, Lyons A, Sherlock G, MacSharry J, Kiely B, Shanahan F, O'Mahony L: Commensal-induced regulatory T cells mediate protection against pathogen-stimulated NF-κB activation. PloS Pathogens 2008;4:e1000112.

40 Wang G, Miyahara Y, Guo Z, Khattar M, Stepkowski SM, Chen W: 'Default' generation of neonatal regulatory T cells. J Immunol 2010;185:71–78.

41 Lee JH, Ulrich B, Cho J, Park J, Kim CH: Progesterone promotes differentiation of human cord blood fetal T cells into T regulatory cells but suppresses their differentiation into Th17 cells. J Immunol 2011; 187:1778–1787.

42 Reading NC, Kasper DL: The starting lineup: key microbial players in intestinal immunity and homeostasis. Front Microbiol 2011;2:148.

P. Brandtzaeg, LIIPAT
Department of Pathology, Rikshospitalet
PO Box 4950
NO–0424 Oslo (Norway)
E-Mail per.brandtzaeg@medisin.uio.no

Guarino A, Quigley EMM, Walker WA (eds): Probiotic Bacteria and Their Effect on Human Health and Well-Being.
World Rev Nutr Diet. Basel, Karger, 2013, vol 107, pp 56–63 (DOI: 10.1159/000345749)

Effects of the Intestinal Microbiota on Behavior and Brain Biochemistry

Eoin Barrett[a–c] · Timothy G. Dinan[a] · John F. Cryan[a] ·
Eamonn M.M. Quigley[a] · Fergus Shanahan[a] · Paul W. O'Toole[a, c] ·
Gerald F. Fitzgerald[a, c] · Catherine Stanton[a, b] · R. Paul Ross[a, b]

[a]Alimentary Pharmabiotic Centre, Biosciences Institute, and [c]Department of Microbiology,
University College Cork, Cork, and [b]Teagasc Food Research Centre, Fermoy, Ireland

Abstract

Increasing evidence is emerging about the influence of the gut microbiota on the brain-gut axis, which has led to the concept of the brain-gut-enteric microbiota axis. In order to study this signaling mechanism, a number of approaches have been used involving animal models which examine the influence of the microbiota on brain function as discussed below. This review focuses on the influence of the gut microbiota on the gut-brain axis, bioactive metabolite production, and the use of probiotics to modulate metabolism and behavior.

The human gastrointestinal (GI) tract harbors trillions of microorganisms which perform vital functions essential for host health, including food processing, digestion of complex indigestible polysaccharides, and synthesis of vitamins [1]. Furthermore, the intestinal microbiota secrete a range of metabolites with functions such as inhibition of pathogens, metabolism of toxic compounds, and production of bioactive compounds [1]. In addition to the aforementioned functions, there is increasing interest in the concept of the 'gut microbiome-gut-brain axis'.

Understanding the influence of the gut microbiota on host health has been described as 'one of the hottest areas in medicine' [2]. For example, a perturbed microbiota has been implicated in an ever-increasing list of disorders in humans including obesity, diabetes, metabolic syndrome, irritable bowel syndrome (IBS), and inflammatory bowel disease in adults [3–5]. As a result, the gut microbiota is increasingly becoming a target for drug and dietary therapy to treat such disorders as well as a source of novel drugs and bioactive compounds [2].

Gut-Brain Axis

The gut-brain axis is a bidirectional signaling between the GI tract and the brain, and is said to be crucial for homeostasis [6]. Behavior and cognitive processes can affect this axis and, indeed, both have been associated with functional disorders such as irritable bowel syndrome (IBS) [6]. It is influenced by neural signals, the hormone corticotrophin-releasing factor, the systemic and mucosal immune systems, and enterochromaffin cells (bidirectional transducers between the gut lumen and the nervous system) [7]. While it was generally accepted that the GI tract microbiota are crucial in immune maturation and gut function, it seemed implausible that the microbiota played a role in brain function [8]. Recently, increasing evidence has emerged about the influence of the gut microbiota on the brain-gut axis, leading to the concept of the brain-gut-enteric microbiota axis; however, the exact mechanisms by which this axis functions remains to be elucidated [6, 9]. This concept is based on the idea that despite the individualistic nature of the gut microbiota in humans, it possesses the ability to communicate with the brain and accordingly modulate host behavior [9]. The brain can communicate with the enteric microbiota directly by releasing signaling molecules into the gut lumen, and indirectly by altering gastric motility and secretion and intestinal permeability [6]. Conversely, the enteric microbiota can communicate with the host via epithelial cells, receptor-mediated signaling, and stimulation of cells of the lamina propria [7]. Furthermore, alterations in the composition of the gut microbiota may lead to deterioration in GI, neuroendocrine, or immune relationships, and these in turn may ultimately lead to disease [9]. In order to study this signaling mechanism, a number of approaches have been used involving various animal models (fig. 1).

Germ-Free Animals
Germ-free animals lack an enteric microbiota and thus are a useful model to study the impact of the GI microbiota on host health and animal behavior [9], and can be used to examine the impact of specific pathogens and probiotics on the central nervous system and the brain-gut axis [8, 9]. Germ-free male animals have been shown to have lower protein levels of brain-derived neurotrophic factor (BDNF), a neurotrophin which impacts on brain function and psychiatric illnesses in the cortex and hippocampus [10]. In addition to decreased serotonin receptor 1A (5HT1A) expression in the hippocampus, an increase in BDNF mRNA expression has been recently reported in the hippocampus of female animals [11]. Despite the discrepancy, the work highlights a link between anxiety and the gut microbiome axis.

Antibiotic/Antimicrobial Intervention
The use of antibiotics/antimicrobials is one of the most common ways of altering the gut microbiota in animals [9]. Bercik et al. [8] administered a mixture of nonabsorbable antimicrobials to specific pathogen-free mice and analyzed behavior, microbiota,

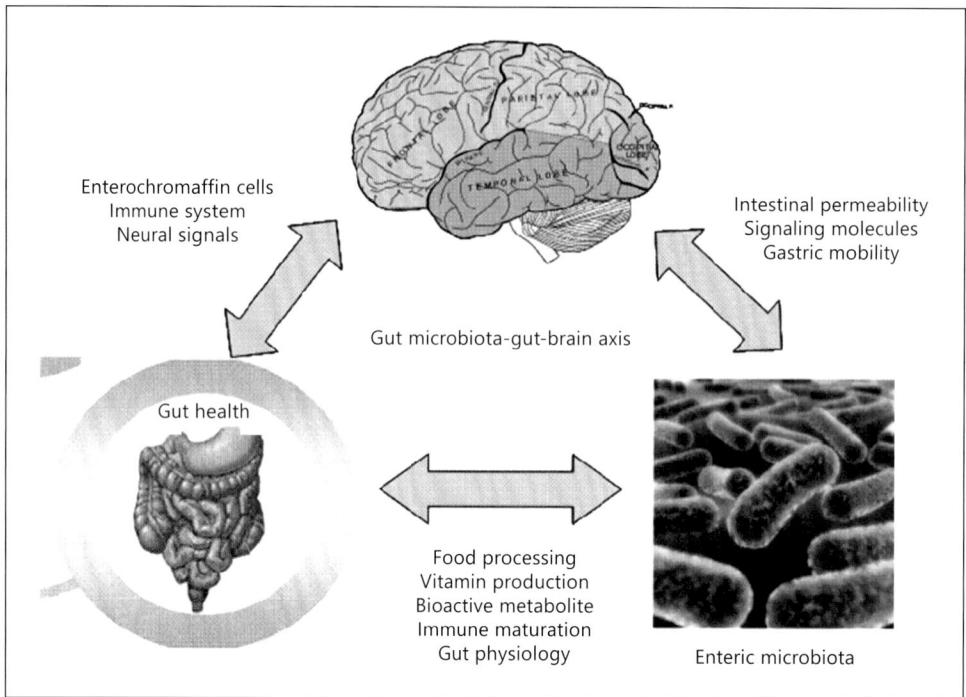

Fig. 1. Pathways of communication and functions of the gut microbiota-gut-brain axis.

cytokines, serotonin, dopamine, and BDNF. Administration of the oral antimicrobials altered the microbiota, increasing Firmicutes and Actinobacteria, increased exploratory behavior, and increased hippocampal expression of BDNF – an effect not seen in germ-free mice [8].

Maternal Separation
Early-life stress has been associated with the development of psychiatric disorders such as depression and anxiety in later life and predisposes individuals to stress-related disorders such as IBS later in life [7]. One such stress is maternal separation, which has been implicated in alterations in the development of the central nervous system [7]. This can be readily modeled in animals using the maternal separation model whereby brief separation of rat pups from the mother triggers long-term changes in colonic sensitivity to rectal distension [7]. Maternally separated animals had increased plasma corticosterone, altered microbiota, behavioral deficits, decreased noradrenaline in the brain, increased visceral sensitivity, and increased immune response; therefore, maternal separation is a valid model for IBS and psychiatric disorders [12]. This animal model has been used to examine the impact of feeding probiotic bacteria on cytokine and tryptophan production, behavior deficits such as forced swim test and colorectal distension, brain noradrenaline concentrations, and γ-aminobutyric acid (GABA) receptor expression [12, 13]. Feeding a number of different probiotic strains

Barrett·Dinan·Cryan·Quigley·Shanahan·O'Toole·Fitzgerald·Stanton·Ross reversed behavioral defects, normalized immune responses, restored noradrenaline concentrations, increased plasma tryptophan levels, and regulated GABA receptor expression in maternally separated animals [12, 13].

Microbial Metabolites

The GI tract microbiota have been shown to produce a range of beneficial metabolites or so-called 'pharmabiotics' which can interact with the host's immune, neural, and endocrine systems and may have a profound effect on health [2].

γ-Aminobutyric Acid
GABA is the main inhibitory neurotransmitter in the brain regulating many physiological and psychological processes, and dysfunctions in the GABA system have been implicated in anxiety and depression [14]. We have recently reported the ability of intestinally derived strains of lactobacilli and bifidobacteria to produce GABA from monosodium glutamate [15]. It has been suggested that microbially produced GABA in the gut may have an effect on the brain-gut axis [16]. Perhaps the ability of *Bifidobacterium* and *Lactobacillus* strains to convert monosodium glutamate to GABA may be considered as a novel probiotic trait, but this remains to be firmly established.

Short-Chain Fatty Acids
Short-chain fatty acids are water-soluble readily absorbable organic fatty acids of 1–6 carbon atoms in length that are found naturally in fruit, vegetables, and milk fats which are produced in the human GI tract as the end-products of anaerobic bacterial fermentation of carbohydrates [17]. Short-chain fatty acids are a major source of energy for colonic epithelial cells, may be a predetermining factor in ulcerative colitis, are involved in prevention of DNA and cell damage, stimulate sodium absorption, and affect immune and inflammatory responses, while sodium butyrate has been shown to elicit an antidepressant effect in the murine brain [18]. Short-chain fatty acids bind to G protein-coupled receptor 43, and this interaction influences inflammatory responses and provides a possible link between GI tract bacterial metabolism and immune and inflammatory responses [19].

Serotonin
Serotonin (5-hyroxytryptophan) is a metabolite of the essential amino acid tryptophan and plays an important role in the regulation of a number of bodily functions, including regulation of GI tract motility, secretion, and sensation. Today, the vast majority of antidepressant drugs lead to increases in the levels of certain neurotransmitters in the brain, in particular norepinephrine and serotonin [20]. Indeed, it has been shown that the plasma serotonin levels of conventional mice

were 2.8-fold higher than germ-free mice [21]. Furthermore, oral ingestion of *Bifidobacterium infantis* increased levels of the serotonin precursor tryptophan in the plasma of rats, suggesting that the strain may have potential as an antidepressant [12].

Endocannabinoids
Endocannabinoids are lipid molecules produced within the body which target receptor sites in the brain [22]. These cell receptor sites also engage with δ^9-tetrahydrocannabinol, the active constituent of *Cannabis sativa* (more commonly known as cannabis), a plant long known for its psychotropic activities [22]. The endocannabinoid system and the gut microbiota can impact on the development of obesity and related disorders [23]. Indeed, changes in the GI tract microbiota selectively decreased endocannabinoid activity in the colon and adipose tissue, a system altered in obesity [23]. In addition, a *Lactobacillus acidophilus* strain modulated expression of cannabinoid receptors and reduced abdominal pain in animals [24].

Probiotics

With the increasing evidence of the potential health benefits of probiotics via the metabolites they produce and the alterations in the microbiota associated with certain disease states, probiotics have emerged as a possible means of alleviating certain disease symptoms. Probiotics are defined as 'live microorganisms which when administered in adequate amounts confer a health benefit on the host' [25].

Probiotics for the Treatment of Irritable Bowel Syndrome
IBS is a poorly understood, widespread, heterogeneous, functional GI tract disorder, characterized by abdominal pain, bloating, altered bowel habits, and discomfort, and is generally viewed as a disorder of the brain-gut axis [26]. The poor success rate of pharmacological interventions to treat this disease has led to a focus on the brain-gut axis to develop therapeutics, which in turn has led to a focus on antibiotic treatment and probiotic intervention [27]. There are numerous studies highlighting the potential of bifidobacteria and lactic acid bacteria to act as therapeutic agents to treat IBS and depression. Suggested mechanisms include altered cytokine and tryptophan production, regulation of cannabinoid expression, normalization of the immune response, reversal of behavior deficits, restoration of basal noradrenaline concentrations, regulation of GABA receptor expression, and modulation of tissue fatty acids [12, 13, 24] (table 1).

Regulation of Central γ-Aminobutyric Acid Receptor Expression
A number of reports have questioned whether GABA production by intestinal bacteria would occur in the gut and if such bioactivity would be beneficial to the host

Barrett · Dinan · Cryan · Quigley · Shanahan · O'Toole · Fitzgerald · Stanton · Ross

Table 1. Suggested mechanisms of action of probiotics as therapeutic agents to treat IBS and depression

Strain	Target	Tissue/organ	Reference
L. rhamnosus	GABA$_{B1b}$ receptor	cortical region, hippocampus, amygdala	13
	GABA$_{A\alpha2}$ receptor	hippocampus, prefrontal cortex, amygdala	13
B. breve	arachidonic acid	brain	28
	docosahexanoic acid	brain	
L. acidophilus	opioid receptor	intestinal epithelial cells	24
	cannabinoid receptor	intestinal epithelial cells	
B. infantis	immune response	blood	12
	noradrenaline	brain	

[16]. Bienenstock et al. [16] were unable to provide a link between pain perception of colorectal distension and GABA levels in the feces of animals fed *Lactobacillus reuteri*. More recently, however, *Lactobacillus rhamnosus* had a direct effect on the GABA receptors in the central nervous system of normal healthy animals [13]. Feeding the strain altered the GABA$_{B1b}$ mRNA in different regions of the mouse brain, as well as GABA$_{A\alpha2}$. Furthermore, administration of the strain reduced stress-induced corticosterone and anxiety- and depression-related behavior. The authors also identified the vagus as a communication pathway between the brain and the bacteria in the gut, with vagotomy preventing the antidepressant effects of the bacterium [13].

Modulation of Fatty Acid Profile
Patients suffering from IBS have elevated plasma levels of arachidonic acid, linked to an increase in proinflammatory eicosanoid production, as well as changes in the species and numbers of the fecal microbiota [27]. The differences in the fatty acid profiles of IBS patients are mirrored in maternally separated rodents. The arachidonic acid content of plasma was significantly elevated in maternally separated animals compared to nonmaternally separated control rodents, suggesting that the fatty acid profiles, in particular the arachidonic acid levels of IBS patients, can be used as markers for IBS [27]. Moreover, modification of the fatty acid profiles of IBS patients through probiotic intervention may be a way to treat the disease. Indeed, it has been shown that altering the gut microbiota changes host fatty acid composition [28]. For example, feeding a *Bifidobacterium* strain can influence the fatty acid composition of tissues, including the colon, liver, and brain [28]. This activity has been found to be strain dependent among different *Bifidobacterium breve* strains. The mechanism by which *B. breve* strains alter the fatty acid composition is uncertain and remains to be elucidated. More recently, eicosapentaenoic acid has been identified as the key ω–3

fatty acid in depression [29]. We have shown that feeding *B. breve* can increase eicosapentaenoic acid in murine tissue [30] and this warrants further investigation for treating depression.

Conclusions

There is increasing evidence that a number of disease states may be linked to disruptions in the brain-gut-microbiome axis. Studies indicate that there is a need to understand the basis for the enteric microbiome-gut-brain axis as alterations in the intestinal microbiota are linked to certain disease states. Furthermore, the microbiota produce a range of metabolites which are beneficial to the host and some of these metabolites may also play a role in the brain-gut axis. Studies focusing on modulation of microbiota composition with pro- and prebiotics as well as exploitation of pharmabiotics that impact the gut-brain axis may provide new therapies, influencing disorders associated with the gut-brain axis, such as IBS, anxiety, and depression.

References

1 Marques TM, Wall R, Ross RP, Fitzgerald GF, Ryan CA, Stanton C: Programming infant gut microbiota: influence of dietary and environmental factors. Curr Opin Biotechnol 2010;21:149–156.

2 Shanahan F: The gut microbiota in 2011: translating the microbiota to medicine. Nat Rev Gastroenterol Hepatol 2012;9:72–74.

3 Ley RE, Turnbaugh PJ, Klein S, Gordon JI: Microbial ecology – human gut microbes associated with obesity. Nature 2006;444:1022–1023.

4 Kassinen A, Krogius-Kurikka L, Makivuokko H, Rinttila T, Paulin L, Corander J, Malinen E, Apajalahti J, Palva A: The fecal microbiota of irritable bowel syndrome patients differs significantly from that of healthy subjects. Gastroenterology 2007;133: 24–33.

5 Peterson DA, Frank DN, Pace NR, Gordon JI: Metagenomic approaches for defining the pathogenesis of inflammatory bowel diseases. Cell Host Microbe 2008;3:417–427.

6 Rhee SH, Pothoulakis C, Mayer EA: Principles and clinical implications of the brain-gut-enteric microbiota axis. Nat Rev Gastroenterol Hepatol 2009;6: 306–314.

7 O'Mahony SM, Hyland NP, Dinan TG, Cryan JF: Maternal separation as a model of brain-gut axis dysfunction. Psychopharmacology 2011;214:71–88.

8 Bercik P, Denou E, Collins J, Jackson W, Lu J, Jury J, Deng Y, Blennerhassett P, Macri J, McCoy KD, Verdu EF, Collins SM: The intestinal microbiota affect central levels of brain-derived neurotropic factor and behavior in mice. Gastroenterology 2011;141: 599–601, 601.e1–e3.

9 Cryan JF, O'Mahony SM: The microbiome-gut-brain axis: from bowel to behavior. Neurogastroenterol Motil 2011;23:187–192.

10 Sudo N, Chida Y, Aiba Y, Sonoda J, Oyama N, Yu XN, Kubo C, Koga Y: Postnatal microbial colonization programs the hypothalamic-pituitary-adrenal system for stress response in mice. J Physiol 2004; 558:263–275.

11 Neufeld KM, Kang N, Bienenstock J, Foster JA: Reduced anxiety-like behavior and central neurochemical change in germ-free mice. Neurogastroenterol Motil 2011;23:255–264, e119.

12 Desbonnet L, Garrett L, Clarke G, Kiely B, Cryan JF, Dinan TG: Effects of the probiotic *Bifidobacterium infantis* in the maternal separation model of depression. Neuroscience 2010;170:1179–1188.

13 Bravo JA, Forsythe P, Chew MV, Escaravage E, Savignac HM, Dinan TG, Bienenstock J, Cryan JF: Ingestion of *Lactobacillus* strain regulates emotional behavior and central GABA receptor expression in a mouse via the vagus nerve. Proc Natl Acad Sci USA 2011;108:16050–16055.

14 Schousboe A, Waagepetersen HS: GABA: Homeostatic and pharmacological aspects; in Tepper JM, Abercrombie ED, Bolam JP (eds): Gaba and the Basal Ganglia: From Molecules to Systems. Amsterdam, Elsevier Science, 2007, pp 9–19.

15 Barrett E, Ross RP, O'Toole PW, Fitzgerald GF, Stanton C: γ-Amino butyric acid production by culturable bacteria from the human intestine. J Appl Microbiol 2012;113:411–417.

16 Bienenstock J, Forsythe P, Karimi K, Kunze W: Neuroimmune aspects of food intake. Int Dairy J 2010; 20:253–258.

17 Cummings JH, Pomare EW, Branch WJ, Naylor CP, Macfarlane GT: Short chain fatty acids in human large intestine, portal, hepatic and venous blood. Gut 1987;28:1221–1227.

18 Schroeder FA, Lin CL, Crusio WE, Akbarian S: Antidepressant-like effects of the histone deacetylase inhibitor, sodium butyrate, in the mouse. Biol Psychiatry 2007;62:55–64.

19 Maslowski KM, Vieira AT, Ng A, Kranich J, Sierro F, Yu D, Schilter HC, Rolph MS, Mackay F, Artis D, Xavier RJ, Teixeira MM, Mackay CR: Regulation of inflammatory responses by gut microbiota and chemoattractant receptor GPR43. Nature 2009;461: 1282–1286.

20 Cryan JF, O'Leary OF: A glutamate pathway to faster-acting antidepressants? Science 2010;329:913–914.

21 Wikoff WR, Anfora AT, Liu J, Schultz PG, Lesley SA, Peters EC, Siuzdak G: Metabolomics analysis reveals large effects of gut microflora on mammalian blood metabolites. Proc Natl Acad Sci USA 2009;106: 3698–3703.

22 Piomelli D: The molecular logic of endocannabinoid signalling. Nat Rev Neurosci 2003;4:873–884.

23 Muccioli GG, Naslain D, Backhed F, Reigstad CS, Lambert DM, Delzenne NM, Cani PD: The endocannabinoid system links gut microbiota to adipogenesis. Mol Syst Biol 2010;6:15.

24 Rousseaux C, Thuru X, Gelot A, Barnich N, Neut C, Dubuquoy L, Dubuquoy C, Merour E, Geboes K, Chamaillard M, Ouwehand A, Leyer G, Carcano D, Colombel JF, Ardid D, Desreumaux P: Lactobacillus acidophilus modulates intestinal pain and induces opioid and cannabinoid receptors. Nat Med 2007;13:35–37.

25 FAO/WHO: Report on Joint FAO/WHO Expert Consultation on Evaluation of Health and Nutritional Properties of Probiotics in Food Including Powder Milk with Live Lactic Acid Bacteria. Córdoba, FAO/WHO, 2001.

26 Drossman DA, Camilleri M, Mayer EA, Whitehead WE: AGA technical review on irritable bowel syndrome. Gastroenterology 2002;123:2108–2131.

27 Clarke G, Quigley EMM, Cryan JF, Dinan TG: Irritable bowel syndrome: towards biomarker identification. Trends Mol Med 2009;15:478–489.

28 Wall R, Marques TM, O'Sullivan O, Ross RP, Shanahan F, Quigley EM, Dinan TG, Kiely B, Fitzgerald GF, Cotter PD, Fuohy F, Stanton C: Contrasting effects of Bifidobacterium breve DPC 6330 and Bifidobacterium breve NCIMB 702258 on fatty acid metabolism and gut microbiota composition. Am J Clin Nutr 2012;95:1278–1287.

29 Martins JG, Bentsen H, Puri BK: Eicosapentaenoic acid appears to be the key omega-3 fatty acid component associated with efficacy in major depressive disorder: a critique of Bloch and Hannestad and updated meta-analysis. Mol Psychiatry 2012, E-pub ahead of print.

30 Wall R, Ross RP, Shanahan F, O'Mahony L, O'Mahony C, Coakley M, Hart O, Lawlor P, Quigley EM, Kiely B, Fitzgerald GF, Stanton C: Metabolic activity of the enteric microbiota influences the fatty acid composition of murine and porcine liver and adipose tissues. Am J Clin Nutr 2009;89:1393–1401.

Catherine Stanton
Teagasc, Moorepark Food Research Centre
Fermoy, Co. Cork (Ireland)
E-Mail catherine.stanton@teagasc.ie

Guarino A, Quigley EMM, Walker WA (eds): Probiotic Bacteria and Their Effect on Human Health and Well-Being.
World Rev Nutr Diet. Basel, Karger, 2013, vol 107, pp 64–71 (DOI: 10.1159/000345735)

Relationship between Bacterial Colonization of Human Digestive and Respiratory Tract

Susan V. Lynch

Division of Gastroenterology, Department of Medicine University of California, San Francisco, Calif., USA

Abstract

The human superorganism represents a coalition of man and microbes, with the greatest diversity and burden of these species concentrated in the lower gastrointestinal (GI) tract. Recent studies in this emerging field have demonstrated relationships between the composition of these communities and diverse aspects of host physiology, including immunological and metabolic function. These studies have implicated GI microbiome dysbiosis as a primary driver of such diseases both within and at sites remote from the GI tract. Epidemiological studies have previously revealed strong correlations between exposures that conceivably impact host microbiota and airway disease. More recent investigations have focused on the impact of such exposures on the GI microbiome and linked these findings to pulmonary outcomes, resulting in a growing body of literature supporting evidence for a gastrointestinal-airway axis. This chapter reviews the recent literature to support this hypothesis and discusses the possibilities for targeting the GI microbiome to improve pulmonary outcomes.

Given the diverse and ubiquitous nature of microbial life, humans have evolved within a proverbial sea of microbial exposure and it is unsurprising that the majority of sites (particularly mucosal surfaces) within and on the human body are now known to be colonized by communities of these organisms [reviewed in 1, 2]. The totality of these microbial assemblages, termed the human microbiome, outnumber host cell numbers at a ratio of 10:1 and represent, particularly in the case of the lower gastrointestinal (GI) microbiome, diverse collections of niche-specific microbial species [3]. Given the potential immunostimulatory capacity of such communities, the host's ability to maintain immune tolerance of these diverse microbial consortia while adeptly discriminating pathogenic from commensal species is stunning. Coevolution of man and its microbiome is, in its simplest manifestation, a symbiosis in which the host provides a protected niche for colonization and the microbiome contributes substantially to a variable array of host-beneficial functions that range from immune development and maintenance of homeostasis [4, 5] to metabolism of indigestible carbohydrates [6]. Indeed the reliance of the mammalian system on microbial colonization is underscored by numerous studies of germ-free mice which demonstrate developmental and functional deficiencies in immune [7], metabolic [8], and neuro-endocrine [9] systems in the absence of microbes.

A large body of evidence points to the birth process as the initial exposure of the nascent human to microbes. Though bacterial species have been detected in amniotic fluid [10, 11], these have largely been associated with preterm or small-for-gestational age babies [12], suggesting that while microbial breach of the amniotic sac is possible, it is not considered typical in healthy full-term pregnancies. The birthing process also appears to dictate the pioneering microbial colonizers of the infant skin and mucosa. Babies born vaginally typically exhibit colonization dominated by *Lactobacillus, Prevotella,* or *Sneathia* species, which are present in the maternal vaginal tract, whereas those born via caesarian section are characterized by maternal skin-associated species including *Staphylococcus, Corynebacterium*, and *Propionibacterium* species [13]. Thus, birth method may largely dictate the pioneer community colonizing the human host, including the GI tract, which has been demonstrated in murine models to play a significant role in educating the developing immune response [7]. Indeed, several studies have examined the development of the neonatal GI microbiome from birth through the initial years of life [14–16]. These investigations have revealed a dramatic increase in bacterial burden over the initial days of life coupled with diversification of the communities (as determined by fecal microbiome profiling), particularly upon the introduction of solid food (typically around 6 months) [14, 16]. These studies demonstrate dramatic perturbation to the bacterial GI assemblages during periods of antibiotic administration. More impressively, they reveal that these highly resilient communities rebound to approximately similar levels of bacterial diversity within a short number of weeks after antibiotic cessation [17]. Nonetheless, specific community members are permanently lost from these assemblages upon such perturbations; whether these losses impact long-term disease development, in particular chronic inflammatory diseases, has yet to be determined.

By convention, studies of the human host are typically performed in a compartmentalized manner, and diseases, even those with system-wide manifestations, e.g. cystic fibrosis, are traditionally studied in discrete compartments, e.g. the airway or the GI tract. However, the emergence of the recent field of human microbiome research has demonstrated evidence for linkage between the GI microbiota and physiological, neurological, and immunological phenomena both within the GI tract and at locations remote from this site [18–20].

The first inklings that a connection exists between GI microbiome composition and airway health is based on a relatively large body of epidemiological literature in the field of childhood asthma development. Asthma is considered a disease of misdirected immune response, typically manifesting as an imbalance in acquired immunity, primarily characterized by activation of a specific population, the Th2 subset of T helper (Th) cells. Th cells are primed by antigen-presenting cells such as dendritic cells that sample the GI microbial milieu and present microbial antigens to T cells, resulting in proliferation and activation of specific Th cell subsets. Central to the dominant theory of perinatal immune programming associated with protection against childhood asthma development is the hygiene hypothesis. In its most recent evolution, the hygiene hypothesis (which perhaps should be termed 'the unhygienic hypothesis') posits that appropriate exposure to environmental microbes results in development of a homeostatic immune system.

That environmental microbial exposure has an impact on allergic disease development is evident from a number of seminal studies of pediatric patients demonstrating that exposure to pets [21] or livestock [22, 23] in infancy results in a significant decrease in the relative risk of childhood asthma development. More recently it has been hypothesized that such exposures afford protection via the microbial species associated with these environments [24, 25]. Moreover, it has been suggested that such microbial exposures influence the composition of the developing infant GI microbiome during a critical period coincident with immune maturation.

Indeed, recent studies have demonstrated that house dust collected from homes with dogs, cats, or no pets differ in their microbial membership and, moreover, that the behavior of pets in the home is associated with distinct house dust microbial exposures [24]. Fujimura et al. [24] recently demonstrated that house dust from homes in which a cat or dog moved freely indoors and outdoors exhibited significantly more diverse bacterial communities and possessed significantly fewer types of fungi. In comparison, house dust from homes with no pets or those in which the animals were maintained exclusively indoors or outdoors, had far fewer types of bacteria and significantly higher fungal species [24]. Thus, pet ownership and behavior of the animal dictates microbial exposures within the home and presumably patterns of GI inoculation and colonization of in-residence infants.

This finding is significant in light of multiple previous studies that demonstrate relationships between early life GI microbiome colonization patterns and allergic disease development in childhood [26–28]. For example, antibiotic use in infancy is associated with a higher risk for asthma development [29], presumably, in light of recent microbiome studies, due to their effects on resident microbiota, particularly in the GI tract. Indeed, a study by Penders et al. [30] showed that an elevated fecal burden of *Escherichia coli* or *Clostridium difficile* at 3 weeks of age was associated with a significantly increased risk of eczema and allergic sensitization at age 6. Furthermore, a recent study by Vael et al. [31] demonstrated that infant fecal colonization at age 3 weeks with either a *Bacteroides fragilis* subgroup or a *C. coccoides* subcluster XIVa species represented an early indicator of asthma development later in life. Moreover, a randomized, double-blind, placebo-controlled study of school children (age: 6–12 years) with asthma and allergic rhinitis who received an oral supplement of either *Lactobacillus gasseri* A5 or placebo daily for 2 months demonstrated significantly increased pulmonary function and decreased clinical symptom scores for asthma and allergic rhinitis in the probiotic-treated patients [32].

Though the implications of microbiome composition in the postnatal GI tract has not yet been determined, recent studies have demonstrated the key role specific bacterial species play in modulating host immune responses and that these immune-activating events may have far-reaching implications beyond the GI tract. For example, mice obtained from two distinct laboratories (Jackson and Taconic labs), though isogenic, exhibit significant differences in the numbers of Th17 cells in the lamina propria of their terminal ileum [33]. Th17 cells, a relatively newly identified subset of Th cells, develop via a distinct (from Th1 and Th2 populations) rho-gamma-mediated pathway, and their overabundance is associated with chronic inflammatory diseases such as inflammatory bowel disease [34], severe asthma [35], and lupus [36]. Comparative microbiome profiling of terminal ileum microbiota of these animals revealed approximately 100 bacterial taxa (groups of bacteria sharing at least 97% sequence similarity in the 16S rRNA gene commonly used for phylogenetic profiling) that were significantly altered in relative abundance.

Among those species exhibiting the greatest magnitude of change in abundance was an unculturable, anaerobic, spore-forming member of the Clostridiales: segmented filamentous bacteria (*Candidatus* Arthromitus). Indeed, mice monocolonized by this specific species illustrated robust induction of Th17 cells, a phenotype not observed when phylogenetically related species were introduced to the animals. Hence, this study illustrates that while a diversity of species inhabit the lower GI tract, it is a small and likely highly specialized subgroup of microbial species that are capable of stimulating distinct T cell subsets. Moreover, it suggests that early life overgrowth of specific bacterial species in this niche putatively leads to aberrant immune programming and inappropriate response to subsequent allergenic challenge at sites within and remote from the GI tract.

Further evidence for the existence of a gastrointestinal-airway axis, and that the GI microbiome may modulate respiratory health status, comes from studies of feeding probiotics to patients with chronic airway disease. Probiotics are defined as live microorganisms which, when administered in adequate amount, confer a health benefit on the host (FAO 2001). In their seminal study of pediatric cystic fibrosis patients, Bruzzese et al. [37] demonstrated in a double-blind, placebo controlled study that children who received daily supplements of *L. rhamnosus* GG exhibited significantly fewer hospitalizations for respiratory infections.

A separate study of pediatric patients fed *L. acidophilus* NCFM, *Bifidobacterium animalis*, or a combination of the two species, demonstrated that children fed probiotic species during the flu season exhibited significantly fewer upper respiratory infections, reduced duration and severity of infection, and fewer absentee days from school [38]. Moreover, children who received both probiotic species demonstrated enhanced protection against infection, indicating that supplementation with multiple commensal species augmented the protective effect [38]. Though the mechanism behind this protective effect is unclear, recent studies implicate both changes in the composition of the GI microbiome and induction of anti-inflammatory T regulatory cell populations.

Cox et al. [39] previously examined the fecal microbiota of a cohort of infants at high risk for asthma, enrolled in the Trial of Infant Probiotic Supplementation (TIPS) study. Infants in this double-blind study received daily supplementation with a probiotic species, *L. rhamnosus* GG, or placebo. Microbiome analyses demonstrated that the presence of a high abundance of this species was associated with substantial restructuring of the GI microbiome. These bacterial communities were characterized by enrichment of multiple commensal species, including several other *Lactobacillus* and a *Bifidobacterium* species, organisms synonymous with health and, more specifically, with protection against allergic disease and childhood asthma development [39].

Murine studies have hinted at the mechanisms associated with probiotic-related protection. Kwon et al. [40] demonstrated that feeding mice a mix of five probiotic species (*L. acidophilus*, *L. casei*, *L. reuteri*, *B. bifidum*, and *Streptococcus thermophiles*) induced populations of CD4+FoxP3+ T regulatory cells. Moreover, they elegantly demonstrated in murine models of inflammatory bowel disease, rheumatoid arthritis, and atopic dermatitis that animals fed this probiotic mix exhibited increased trafficking of T reg cells to the affected site and significant abrogation of inflammation [40]. Thus, the investigators demonstrated that the effect of commensal species supplementation on host immune responses was not simply restricted locally to the GI tract, but evident at remote sites of modeled inflammatory disease. In a subsequent review by Rauch and Lynch [41], the authors noted that the species used to supplement mice in these studies were also among those species most significantly increased in relative abundance in infants supplemented with *Lactobacillus* GG (TIPS trial described above), suggesting that a similar mechanism involving commensal induction of anti-inflammatory T reg cells may indeed protect against asthma development.

Though no studies to date have been performed in the airways of pediatric asthmatic populations, recent investigations of adult asthmatics have clearly demonstrated the presence of a diverse airway microbiome [42, 43]. However, published studies to date have examined patients using inhaled corticosteroids, thus it remains unclear whether the detected airway diversity is due to asthma or the local immunosuppression afforded by steroid use. The latter, due to the selective effects of corticosteroids on microbial species, is most likely. A recent study demonstrating, albeit in a very small cohort of patients, that severely immunosuppressed HIV-infected patients with pneumonia exhibit substantially greater airway diversity compared to comparable non-HIV-infected pneumonia patients supports this notion [44].

Despite the inevitable caveat of corticosteroid use, Huang et al. [42] demonstrated that asthmatic patients exhibit significantly greater bacterial burden and that the degree of bronchial hyperresponsiveness is strongly associated with the diversity within the resident microbiome, i.e. greater bacterial diversity is associated with more responsive airways. The findings from this study were also concordant with an earlier airway microbiome study performed by Hilty et al. [43], who demonstrated enrichment of Proteobacteria in the airways of asthmatic patients (compared to control subjects).

Though a link between GI microbiome composition and airway disease in asthmatic adults has yet to be determined, the increasing prevalence of this disease specifically in Western nations is noteworthy. Very recent studies have demonstrated that long-term dietary habits are associated with specific and distinct GI microbiota; Western diet consumption results in a *Bacteroides*-centered consortia distinct from that of individuals who consume a largely plant-based diet and are characterized by consortia enriched for *Prevotella* [45, 46]. The health implications of possessing such distinct communities are far from clear; however, the observation that inflammatory and autoimmune disease burden has increased dramatically in Western nations and that the associated dramatic shift in diet and lifestyle radically reshapes GI microbiome composition implicates these consortia in this phenomenon.

Although the field of human microbiome research is in the proverbial neonatal stage, the implications for human health are already staggering. The conglomerate human-microbial superorganism represents a highly integrated and complex system, influenced by its environment, diet, and host genetics. Perturbations to these components either through industrialization (loss of microbial exposure), diet and lifestyle, or antimicrobial use, appear to have dramatic implications for human health. That the GI microbiome represents a critical component in host health is evident from a large number of studies that have demonstrated microbial dysbiosis in this niche associated with chronic inflammatory and autoimmune diseases [47, 48]. What is more stunning is that the composition of microbes at this site appears to impact inflammatory status at sites remote form the GI tract, including the respiratory tract. This opens up the possibility for novel therapeutic strategies involving GI microbiome restoration ecology approaches, much like fecal transplantation, but perhaps using a reduced and more refined consortium of microbes tailored to specific individuals and their resident bacterial assemblages. Such approaches are likely to result in dramatic improvements in outcomes and represent an alternative and highly efficacious avenue for treatment of both GI and airway diseases.

References

1 Grice EA, Segre JA: The skin microbiome. Nat Rev Microbiol 2011;9:244–253.
2 Walter J, Ley R: The human gut microbiome: ecology and recent evolutionary changes. Annu Rev Microbiol 2011;65:411–429.
3 Ley RE, Peterson DA, Gordon JI: Ecological and evolutionary forces shaping microbial diversity in the human intestine. Cell 2006;124:837–848.
4 Taschuk R, Griebel PJ: Commensal microbiome effects on mucosal immune system development in the ruminant gastrointestinal tract. Anim Health Res Rev 2012;13:129–141.
5 Kelly D, Mulder IE: Microbiome and immunological interactions. Nutr Rev 2012;70(suppl 1):S18–S30.
6 Scott KP, Duncan SH, Louis P, Flint HJ: Nutritional influences on the gut microbiota and the consequences for gastrointestinal health. Biochem Soc Trans 2011;39:1073–1078.
7 Cebra JJ, Periwal SB, Lee G, Lee F, Shroff KE: Development and maintenance of the gut-associated lymphoid tissue (GALT): the roles of enteric bacteria and viruses. Dev Immunol 1998;6:13–18.
8 Backhed F: Programming of host metabolism by the gut microbiota. Ann Nutr Metab 2011;58(suppl 2):44–52.

9 Neufeld KM, Kang N, Bienenstock J, Foster JA: Reduced anxiety-like behavior and central neurochemical change in germ-free mice. Neurogastroenterol Motil 2011;23:255–264.

10 DiGiulio DB, et al: Prevalence and diversity of microbes in the amniotic fluid, the fetal inflammatory response, and pregnancy outcome in women with preterm pre-labor rupture of membranes. Am J Reprod Immunol 2010;64:38–57.

11 DiGiulio DB, et al: Microbial prevalence, diversity and abundance in amniotic fluid during preterm labor: a molecular and culture-based investigation. PLoS One 2008;3:e3056.

12 DiGiulio DB, et al: Microbial invasion of the amniotic cavity in pregnancies with small-for-gestational-age fetuses. J Perinat Med 2010;38:495–502.

13 Dominguez-Bello MG, et al: Delivery mode shapes the acquisition and structure of the initial microbiota across multiple body habitats in newborns. Proc Natl Acad Sci USA 2010;107:11971–11975.

14 Koenig JE, et al: Succession of microbial consortia in the developing infant gut microbiome. Proc Natl Acad Sci USA 2011;108(suppl 1):4578–4585.

15 Morowitz MJ, et al: Strain-resolved community genomic analysis of gut microbial colonization in a premature infant. Proc Natl Acad Sci USA 2011;108: 1128–1133.

16 Palmer C, Bik EM, DiGiulio DB, Relman DA, Brown PO: Development of the human infant intestinal microbiota. PLoS Biol 2007;5:e177.

17 Dethlefsen L, Huse S, Sogin ML, Relman DA: The pervasive effects of an antibiotic on the human gut microbiota, as revealed by deep 16S rRNA sequencing. PLoS Bio 2008;6:e280.

18 Round JL, O'Connell RM, Mazmanian SK: Coordination of tolerogenic immune responses by the commensal microbiota. J Autoimmun 2010;34:J220–J225.

19 Iebba V, Aloi M, Civitelli F, Cucchiara S: Gut microbiota and pediatric disease. Dig Dis 2011;29:531–539.

20 Clemente JC, Ursell LK, Parfrey LW, Knight R: The impact of the gut microbiota on human health: an integrative view. Cell 2012;148:1258–1270.

21 Aichbhaumik N, et al: Prenatal exposure to household pets influences fetal immunoglobulin E production. Clin Exp Allergy 2008;38:1787–1794.

22 von Mutius E, Vercelli D: Farm living: effects on childhood asthma and allergy. Nat Rev Immunol 2010;10:861–868.

23 von Mutius E: Asthma and allergies in rural areas of Europe. Proc Am Thorac Soc 2007;4:212–216.

24 Fujimura KE, et al: Man's best friend? The effect of pet ownership on house dust microbial communities. J Allergy Clin Immunol 2010;126:410–412.

25 von Mutius E: 99th Dahlem conference on infection, inflammation and chronic inflammatory disorders: farm lifestyles and the hygiene hypothesis. Clin Exp Immunol 2010;160:130–135.

26 van Nimwegen FA, et al: Mode and place of delivery, gastrointestinal microbiota, and their influence on asthma and atopy. J Allergy Clin Immunol 2011;128: 948–955.

27 Penders J, Stobberingh EE, van den Brandt PA, Thijs C: The role of the intestinal microbiota in the development of atopic disorders. Allergy 2007;62: 1223–1236.

28 Kalliomaki M, Isolauri E: Pandemic of atopic diseases – a lack of microbial exposure in early infancy? Curr Drug Targets Infect Disord 2002;2:193–199.

29 Johnson CC, et al: Antibiotic exposure in early infancy and risk for childhood atopy. J Allergy Clin Immunol 2005;115:1218–1224.

30 Penders J, et al: Gut microbiota composition and development of atopic manifestations in infancy: the KOALA Birth Cohort Study. Gut 2007;56:661–667.

31 Vael C, Vanheirstraeten L, Desager KN, Goossens H: Denaturing gradient gel electrophoresis of neonatal intestinal microbiota in relation to the development of asthma. BMC Microbiol 2011;11:68.

32 Chen YS, Jan RL, Lin YL, Chen HH, Wang JY: Randomized placebo-controlled trial of Lactobacillus on asthmatic children with allergic rhinitis. Pediatr Pulmonol 2010;45:1111–1120.

33 Ivanov II, et al: Induction of intestinal Th17 cells by segmented filamentous bacteria. Cell 2009;139:485–498.

34 Geremia A, Jewell DP: The IL-23/IL-17 pathway in inflammatory bowel disease. Expert Rev Gastroenterol Hepatol 2012;6:223–237.

35 Aujla SJ, Alcorn JF: T(H)17 cells in asthma and inflammation. Biochim Biophys Acta 2011;1810: 1066–1079.

36 Chen DY, et al: The potential role of Th17 cells and Th17-related cytokines in the pathogenesis of lupus nephritis. Lupus 2012;21:1385–1396.

37 Bruzzese E, et al: Effect of Lactobacillus GG supplementation on pulmonary exacerbations in patients with cystic fibrosis: a pilot study. Clin Nutr 2007;26: 322–328.

38 Leyer GJ, Li S, Mubasher ME, Reifer C, Ouwehand AC: Probiotic effects on cold and influenza-like symptom incidence and duration in children. Pediatrics 2009;124:e172–e179.

39 Cox MJ, et al: Lactobacillus casei abundance is associated with profound shifts in the infant gut microbiome. PLoS One 2010;5:e8745.

40 Kwon HK, et al: Generation of regulatory dendritic cells and CD4+Foxp3+ T cells by probiotics administration suppresses immune disorders. Proc Natl Acad Sci USA 2010;107:2159–2164.

41 Rauch M, Lynch SV: The potential for probiotic manipulation of the gastrointestinal microbiome. Curr Opin Biotechnol 2012;23:192–201.

42 Huang YJ, et al: Airway microbiota and bronchial hyperresponsiveness in patients with suboptimally controlled asthma. J Allergy Clin Immunol 2011; 127:372–381.

43 Hilty M, et al: Disordered microbial communities in asthmatic airways. PLoS One 2010;5:e8578.

44 Iwai S, et al: Oral and airway microbiota in HIV-infected pneumonia patients. J Clin Microbiol 2012;50: 2995–3002.

45 Wu GD, et al: Linking long-term dietary patterns with gut microbial enterotypes. Science 2011;334: 105–108.

46 Arumugam M, et al: Enterotypes of the human gut microbiome. Nature 2011;473:174–180.

47 Tlaskalova-Hogenova H, et al: The role of gut microbiota (commensal bacteria) and the mucosal barrier in the pathogenesis of inflammatory and autoimmune diseases and cancer: contribution of germ-free and gnotobiotic animal models of human diseases. Cell Mol Immunol 2011;8:110–120.

48 Hooper LV, Littman DR, Macpherson AJ: Interactions between the microbiota and the immune system. Science 2012;336:1268–1273.

Susan V. Lynch
Division of Gastroenterology, Department of Medicine University of California
513 Parnassus Ave.
San Francisco, CA 94143-0538 (USA)
E-Mail Susan.lynch@ucsf.edu

Guarino A, Quigley EMM, Walker WA (eds): Probiotic Bacteria and Their Effect on Human Health and Well-Being.
World Rev Nutr Diet. Basel, Karger, 2013, vol 107, pp 72–78 (DOI: 10.1159/000345736)

Probiotics in the Prevention and Treatment of Inflammatory Bowel Diseases in Children

Bénédicte Pigneur[a, b] · Frank M. Ruemmele[a, b]

[a]Université Paris Descartes, Sorbonne Paris Cité, and [b]Assistance Publique-Hôpitaux de Paris,
Hôpital Necker-Enfants Malades, Service de Gastroentérologie pédiatrique, Paris, France

Abstract

Inflammatory bowel diseases (IBD) are an aberrant immune reaction most likely against the intestinal microbiota in genetically determined high-risk individuals. Several recent studies have highlighted that the intestinal microbiome is markedly disturbed in patients with IBD. This pathological state is now called dysbiosis. It is therefore reasonable to develop strategies to restore the altered microbiome in children with IBD. However, few clinical trials in children have been published, and to date only few approaches have entered into clinical practice. These trials, as well as the concept of using living bacteria (probiotics) to treat IBD, are reviewed and discussed in detail in the present review.

Copyright © 2013 S. Karger AG, Basel

Inflammatory bowel diseases (IBD) are characterized by chronic inflammation of the intestinal tract of unknown origin. While inflammation in ulcerative colitis (UC) is restricted to the rectum and colon, Crohn's disease (CD) can affect the entire gastrointestinal tract from the mouth to the anus. Disease onset is usually during young adulthood (the third decade of life), but in about 20% of patients IBD starts under the age of 18 years [1]. Since the 1950s, the incidence of IBD, and especially CD, has been increasing in developed countries without a clear explanation. While a plateau was reached in the late 1990s in adult IBD populations, the incidence of IBD in children is still increasing, particularly in young children below the age of 10 years [2].

Although major progress has been achieved in recent years, the pathogenesis of IBD has not been fully elucidated. It is well accepted that IBD reflect chronic intestinal disorders with an aberrant immune response towards luminal antigens, most probably commensal bacteria in genetically susceptible subjects. With the development of molecular techniques making bacterial cultures unnecessary, the study of intestinal microbiota has become more accurate and precise in recent years. Understanding the mechanisms of action of intestinal bacteria and their interaction with host cells represents a major challenge for understanding IBD, as well as other inflammatory diseases.

Study of the Microflora

The intestinal microbiota is composed of all bacteria living in the digestive tract. Humans harbor approximately 10^{14} bacteria in their digestive tract, which outnumbers the number of cells of the total human body (10^{13} eukaryotic cells). The microbiota represents a huge biomass with important and useful functions to the host. Only 10–30% of fecal bacteria are cultivable. Access to the study of noncultivable bacteria was made possible by molecular methods, based on the analysis of the gene for bacterial 16S ribosomal DNA. This gene encodes the ubiquitous RNA small subunit of the bacterial ribosome. The study of conserved regions (common to all of the Bacteria domain) and variables (common to a group of bacteria) is used to classify bacteria phylogenetically within the same group and to determine in a fecal sample all of the different bacterial species.

At birth, the gastrointestinal tract is sterile. Initial exposure of the gut to microbes occurs during delivery from the maternal fecal and vaginal flora. Within months after birth, a relatively stable microbial population is established. This abundant, diverse, and dynamic intestinal microflora normally lives in a complex symbiotic relationship with the eukaryotic cells of the mucosa. The fecal microbiota of an individual has hundreds of different species belonging to about 50 genera. Analysis of its composition in major phylogenetic groups revealed the existence of recurring components, found in all individuals. Three bacterial phyla, Firmicutes, Bacteroidetes, and Actinobacteria, together represent the largest dominant fecal bacteria (fig. 1).

Rationale for the Use of Probiotics in Inflammatory Bowel Diseases

Clinical Studies

The intestinal microflora is an important element in the induction and chronicity of lesions in CD. Inflammatory lesions in IBD are described preferentially at the distal ileum and colon, precisely where the concentrations of microorganisms are the highest in the digestive tract [3]. Numerous studies have shown that the concentrations of bacteria associated with the mucosa were higher in patients with IBD than in controls [4].

In humans, the most convincing clinical evidence are observed in the postoperative recurrence of CD in patients with ileostomy; relapse is observed only if the fecal stream is maintained on the anastomosis or if all of the ileal liquid is instilled into the colon, but not if a filtrate of ileal fluid is used. Rutgeerts et al. [5] showed that

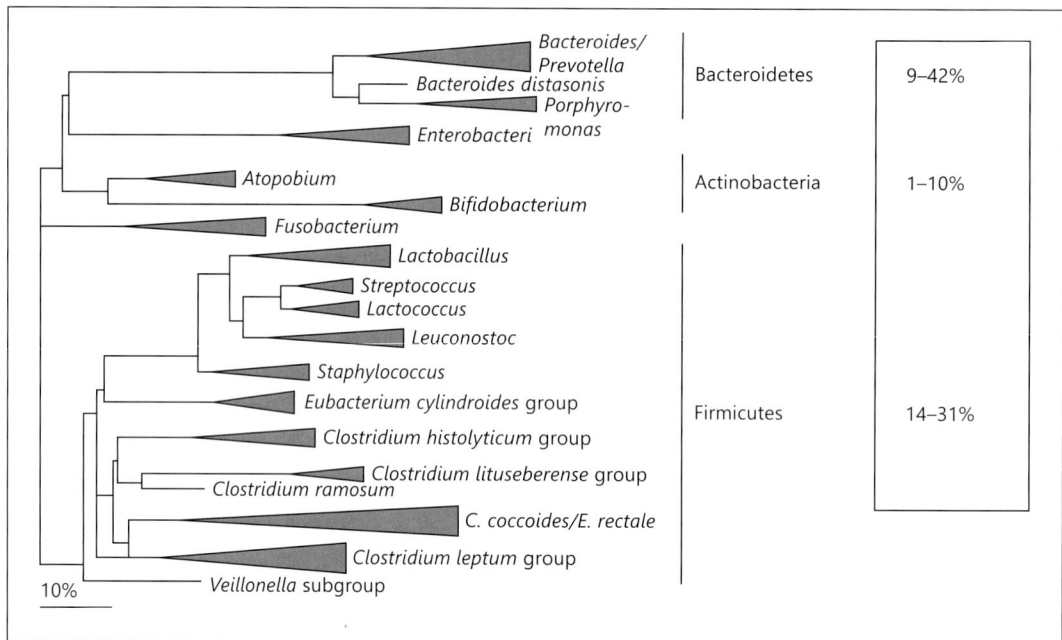

Fig. 1. Distribution of the three main phyla (bacteria of the dominant flora) expressed in percentage of total bacteria.

after resection of the terminal ileum, recurrence of CD occurred in 70% of patients within 6 months after the restoration of intestinal continuity, but was not observed in patients with permanent ileostomy. Moreover, some antibiotics, particularly metronidazole, have established efficacy in preventing postoperative recurrence of CD.

Experimental Models

The role of bacteria in the pathogenesis of IBD is also supported by studies in animals. The development of colitis in animal models of spontaneous colitis (HLA-B27 transgenic rats, knockout mice for IL-10, IL-2, or mice overexpressing TNF) or induced by chemical agents (dextran sulfate sodium, acetic acid, and sodium trinitrobenzosulfonate) requires the presence of a bacterial microflora in the intestinal lumen [6]. Inflammation is absent in animals maintained in axenic conditions, and symptoms appear when a digestive microflora develops.

Genetics

Most genes known to predispose to IBD (NOD2, IL-23R, ATG16L1, and IRGM) are involved in interactions between bacteria and the immune system. The genetic association most clearly established is that of NOD2 with ileal CD. NOD2 codes for an intracellular receptor allowing the cells to detect the presence of bacteria in the intracellular compartment, and is activated by a component of bacterial peptidoglycan, the muramyl dipeptide, present on Gram-negative and Gram-positive bacteria [7]. A meta-analysis showed an odds ratio of 2.4 if the patient is heterozygous and 17.1 for homozygotes [8]. This genetic polymorphism supports the hypothesis that CD results from an inappropriate immune response to intestinal microflora in a genetically predisposed subject.

Dysbiosis and Inflammatory Bowel Diseases

Several studies, based mainly on bacterial cultures, showed that the fecal microflora of IBD patients was different from that of control subjects and described an increase of Enterobacteriaceae during relapse in CD [9, 10]. More recently, studies based on the recognition of the 16S ribosomal RNA molecule and/or their genes allowed for a more precise description of the composition of the microflora in IBD patients. Sokol et al. [11] showed a significant decrease in the proportion of the *Clostridium leptum* phylogenetic group in patients with colonic CD. These results were confirmed by a metagenomic approach revealing a restriction in biodiversity depending on the bacteria belonging to Firmicutes (*C. leptum* and *C. coccoides* groups) with a decrease in the proportion of bacteria belonging to the *C. leptum* phylogenetic group [12, 13]. These elements proved the existence of an imbalance of the microbiota in IBD, now named 'dysbiosis'. This dysbiosis is characterized by high instability of the microbiota over time, the presence of about 30% of unusual bacteria, an increase in mucosal bacterial concentration, and a restriction of biodiversity regarding the Firmicutes phylum [14]. Within this group, a decrease in the *C. leptum* group has been shown and particularly of its major representative, *Faecalibacterium prausnitzii*. In a pediatric population of CD, Schwiertz et al. [15] showed a decrease in *F. prausnitzii* and an increase in *Escherichia coli*. The loss of *F. prausnitzii* associated with the development of IBD suggests an anti-inflammatory role of this bacteria. This has been verified on cellular and animal models [16]. The identification of the molecules responsible for this anti-inflammatory activity is the subject of extensive research.

Probiotics in Inflammatory Bowel Diseases

From a therapeutic standpoint, it appears logical to correct the dysbiosis observed in IBD patients with a probiotic approach. Probiotics are live bacteria with a proven beneficial health effect on ingestion. Many different probiotic bacteria have been tested as treatment options for inflammatory diseases, largely supported by animal studies. For instance, the exogenous administration of *Lactobacillus reuteri* prevented the development of experimental colitis induced by acetic acid or methotrexate in rats [17].

Data in humans suggest that probiotics are capable of preventing relapse of chronic intestinal inflammation. But there is a relative lack of rigorously designed randomized placebo-controlled trials. The majority of studies have been conducted in adult IBD patients and only a few have included children with IBD.

The most convincing trials with probiotics to treat IBD were in the context of pouchitis in adults. A positive effect of the administration of probiotics, a mixture of eight different strains (VSL#3), was observed in this particular case [18].

A controlled study showed that the probiotic *E. coli* Nissle 1917 is as effective as oral mesalazine in the maintenance of remission in adult UC patients [19]. Huynh et al. [20] evaluated the efficacy of VSL#3 in the treatment of mild-to-moderate UC in children and reported a remission rate of 56% after 8 weeks of treatment. Another pediatric, randomized, placebo-controlled trial suggested the efficacy and safety of VSL#3 in active UC and demonstrated its role in maintenance of remission [21].

In CD, evidence for the efficacy of probiotics for the prevention of relapse are few and mostly contradictory, except in children where the administration of *Lactobacillus* GG (LGG) showed a decreased activity of illness. In children, the most widely used probiotic is LGG. The addition of LGG to prednisone decreased disease activity in a small study in children [9]; however, in a larger study in children there was no difference in remission rate observed over 2 years [22].

There are multiple mechanisms of action in probiotics: modulation/re-equilibration of the intestinal bacterial flora, modification of the mucus production reinforcing the intestinal epithelial barrier, production and secretion of anti-inflammatory molecules, stimulation of repair mechanisms of the epithelium, and increased synthesis of antibacterial peptides by enterocytes, Paneth cells, etc. Recently, it was shown in animal models that the subcutaneous injection of *L. salivarius* had an anti-inflammatory effect, indicating a systemic effect [23]. This could be achieved by the immunostimulatory effect by DNA sequences of probiotics via Toll-like receptor 9 [24]. It was also shown that probiotics are capable of stimulating repair mechanisms of the epithelium and increasing the synthesis of antibacterial peptides.

Recombinant Bacteria

The creation of genetically modified probiotics for the production of new proteins offers new therapeutic approaches. It could be considered as an original way to deliver active constituents to targets in the gastrointestinal tract. This approach would allow (1) bringing the protein directly into the intestinal ecosystem, (2) a possible secretion of this protein by the microorganism, (3) the conditions necessary for the secretion, and (4) to obtain long-term effects if the probiotic is able to colonize the ecological niche. There is one example for this theoretically very convincing approach: Steidler et al. [25] have modified the genome of a *Lactococcus lactis* strain to make it produce high levels of IL-10. Intragastric administration of this IL-10-secreting probiotic induced a 50% reduction of colitis in mice with dextran sulfate/induced colitis and it prevented the onset of colitis in IL-10 knockout mice. Other attempts to translate these findings to patients with IBD are underway [26]. Therefore, probiotic vectors could provide an original new way to deliver active ingredients at the site of inflammation.

In conclusion, the microbiota appears to be a key element in the pathogenesis of IBD. Dysbiosis reflects an imbalance of 'beneficial' and 'harmful' bacteria in the host and could participate in an inappropriate and chronic activation of the intestinal immune system thereby leading to a chronic inflammatory process. Although the therapeutic use of probiotics to treat pediatric forms of IBD is so far rather disappointing, the idea to correct dysbiosis in these patients still remains very convincing. It is remarkable to observe a tendency in clinical trials indicating that adult IBD patients respond slightly better to probiotics compared to children. Whether this points to a different sensitivity toward probiotics between adults and children has to be confirmed. The diverging results might simply reflect different clinical designs since there is no convincing rationale to believe that children are less responsive to probiotics than adults; on the contrary, there are many arguments to expect that children are even more responsive.

References

1 Ruemmele FM: Pediatric Crohn's disease – coming of age? Curr Opin Gastroenterol 2010;26:332–336.

2 Benchimol EI, Fortinsky KJ, Gozdyra P, et al: Epidemiology of pediatric inflammatory bowel disease: a systematic review of international trends. Inflamm Bowel Dis 2011;17:423–439.

3 Malaty HM, Fan X, Opekun AR, et al: Rising incidence of inflammatory bowel disease among children: a 12-year study. J Pediatr Gastroenterol Nutr 2010;50:27–31.

4 Swidsinki A, Ladhoff A, Pernthaler A, et al: Mucosal flora in inflammatory bowel disease. Gastroenterology 2002;122:44–54.

5 Rutgeerts P, Goboes K, Peeters M, et al: Effect of faecal stream diversion on recurrence of Crohn's disease in the neoterminal ileum. Lancet 1999;338:771–774.

6 Elson CO, Sartor RB, Tennyson GS, et al: Experimental models of inflammatory bowel disease. Gastroenterology 1995;109:1344–1367.

7 Hugot JP, Chamaillard M, Zouali H, et al: Association of NOD2 leucine-rich repeat variants with susceptibility to Crohn's disease. Nature 2001;411:599–603.

8 Kobayashi KS, Chamaillard M, Ogura Y, et al: Nod2-dependent regulation of innate and adaptative immunity in the intestin tract. Science 2005;307:731–734.

9 Guandalini S: Use of Lactobacillus-GG in paediatric Crohn's disease. Dig Liver Dis 2002;34(suppl 2):S63–S65.

10 Langlands SJ, Hopkins MJ, Coleman M, et al: Prebiotic carbohydrates modify the mucosa associated microflora of the human large bowel. Gut 2004;53:1610–1616.

11 Sokol H, Seksik P, Furet JP, et al: Low counts of *Faecalibacterium prausnitzii* in colitis microbiota. Inflamm Bowel Dis 2009;15:1183–1189.

12 Manichanh C, Rigottier-Gois L, Bonnaud E, et al: Reduced diversity of faecal microbiota in Crohn's disease revealed by a metagenomic approach. Gut 2006;55:205–211.

13 Mangin I, Bonnet R, Seksik P, et al: Molecular inventory of faecal microflora in patients with Crohn's disease. FEMS Microbiol Ecol 2004;50:25–36.

14 Seksik P: Microbiote intestinal et MICI. Gastroentérologie Clinique Biologique 2010;34:48–55.

15 Schwiertz A, Jacobi M, Frick JS, et al: Microbiota in pediatric inflammatory bowel disease. J Pediatr 2010;157:240–244.

16 Sokol H, Pigneur B, Watterlot L, et al: *Faecalibacterium prausnitzii* is an anti-inflammatory commensal bacterium identified by gut microbiota analysis of Crohn disease patients. Proc Natl Acad Sci USA 2008;105:16731–16736.

17 Mao Y, Nobaek S, Kasravi B, et al: The effects of Lactobacillus strains and oat fiber on methotrexate-induced enterocolitis in rats. Gastroenterology 1996;111:334–344.

18 Gionchetti P, Rizzello F, Helwig U, et al: Prophylaxis of pouchitis onset with probiotic therapy: a double-blind, placebo-controlled trial. Gastroenterology 2003;124:1202–1209.

19 Kruis W, Fric P, Pokrotnieks J, et al: Maintaining remission of ulcerative colitis with the probiotic *Escherichia coli* Nissle 1917 is as effective as with standard mesalazine. Gut 2004;53:1617–1623.

20 Huynh HQ, De Bruyn J, Guan L, et al: Probiotic preparation VSL#3 induces remission in children with mild to moderate acute ulcerative colitis: a pilot study. Inflamm Bowel Dis 2009;15:760–768.

21 Miele E, Pascarella F, Giannetti E, et al: Effect of a probiotic preparation (VSL#3) on induction and maintenance of remission in children with ulcerative colitis. Am J Gastroenterol 2009;104:437–443.

22 Bousvaros A, Guandalini S, Baldassano RN, et al: A randomized, double-blind trial of Lactobacillus GG versus placebo in addition to standard maintenance therapy for children with Crohn's disease. Inflamm Bowel Dis 2005;11:883–889

23 Sheil B, McCarthy J, O'Mahony L, et al: Is the mucosal route of administration essential for probiotic function? Subcutaneous administration is associated with attenuation of murine colitis and arthritis. Gut 2004 53:694–700.

24 Rachmilewitz D, Katakura K, Karmeli F, et al: Toll-like receptor 9 signaling mediates the anti-inflammatory effects of probiotics in murine experimental colitis. Gastroenterology 2004;126:520–528.

25 Steidler L, Hans W, Schotte L, et al: Treatment of murine colitis by *Lactococcus lactis* secreting interleukin-10. Science 2000;289:1352–1355.

26 Braat H, Rottiers P, Hommes DW, et al: A phase I trial with transgenic bacteria expressing interleukin-10 in Crohn's disease. Clin Gastroenterol Hepatol 2006;4:754–759.

Frank M. Ruemmele
Assistance Publique-Hôpitaux de Paris, Hôpital Necker-Enfants Malades
Service de Gastroentérologie pédiatrique
149 rue de Sèvres, FR–75015 Paris (France)
E-Mail frank.ruemmele@nck.aphp.fr

Probiotics in the Prevention and Treatment of Diseases in Adults and Children

Guarino A, Quigley EMM, Walker WA (eds): Probiotic Bacteria and Their Effect on Human Health and Well-Being.
World Rev Nutr Diet. Basel, Karger, 2013, vol 107, pp 79–86 (DOI: 10.1159/000345737)

Functional Gastrointestinal Disorders in Children

Flavia Indrio[a] · Giuseppe Riezzo[b]

[a]Department of Pediatrics, University of Bari, Bari, [b]Laboratory of Experimental Pathophysiology,
IRCCS 'S de Bellis', Castellana Grotte (BA), Italy

Abstract

Functional gastrointestinal disorders are a complex of clinical entities characterized by disorder in function at the level of the gastrointestinal tract or in the central processing of information originating in the gastrointestinal tract. The etiology is multifactorial. Alterations of gut motility, visceral hypersensitivity, neural function, intestinal inflammation without anatomical lesion, and the gut-brain axis are implicated in the disease. Recently, an additional etiological factor was investigated. An abnormal intestinal microbiota in children might influence intestinal barrier function, activate innate intestinal immune response, dysregulate the enteric nervous system, and alter visceral sensitivity and intestinal motility. The possibility to target probiotics as a new therapeutic approach in this condition is also discussed.

Copyright © 2013 S. Karger AG, Basel

Functional gastrointestinal disorders (FGIDs) are defined as a variable combination of chronic or recurrent gastrointestinal symptoms not explained by structural or biochemical abnormalities [1], the primary abnormality being an alteration of physiological function at the level of the gastrointestinal tract or in the central processing of information originating in the gastrointestinal tract.

In childhood, FGIDs often vary because of the developmental stage of the child. The Rome III process established two pediatric committees. The infant/toddler (up to 4 years) committee [2] and the child/adolescent (4–18 years) committee focus on the criteria for FGIDs (table 1) [3]. Several factors are involved, so the expression of an FGID depends on an individual's autonomic, affective, and intellectual developmental stage. This makes the pathophysiology underlying these disorders even more complicated [4].

Visceral hyperalgesia is thought to be the first step in the pathogenesis of FGIDs. Genetically predisposed individuals undergoing a wide array of sensitizing events can

Table 1. Rome III diagnostic criteria (appendix A) for FGIDs: sections G and H

G		Childhood FGIDs: infant/toddler
1		Infant regurgitation
2		Infant rumination syndrome
3		Cyclic vomiting syndrome
4		Infant colic
5		Functional diarrhea
6		Infant dyschezia
7		Functional constipation
H		Childhood FGIDs: child/adolescent
1		Vomiting and aerophagia
	a	Adolescent rumination syndrome
	b	Cyclic vomiting syndrome
	c	Aerophagia
2		Abdominal pain-related functional gastrointestinal disorders
	a	Functional dyspepsia
	b	IBS
	c	Abdominal migraine
	d	Childhood functional abdominal pain
	d1	Childhood functional abdominal pain syndrome
3		Constipation and incontinence
	a	Functional constipation
	b	Nonretentive fecal incontinence

develop visceral hyperalgesia. Environmental factors that may act as precipitating events, including intestinal infections, allergies, distention, and all stressors that exceed the child's coping abilities, have the potential to disturb the homeostasis of the brain-gut axis, eventually resulting in chronic or recurrent abdominal pain. Sensitizing psychosocial events, such as mood disorders and anxiety, the child's appraising and coping style, the family's coping style, psychological morbidities in the mother, and the presence of potential secondary gains for the patient may lead to loss of function and increased disability [5, 6].

The role of early life events in causing FGIDs later in life is still controversial, even though multifunction neuromotor ions at birth (i.e. gastric suction) can promote the development of long-term visceral hypersensitivity and cognitive hypervigilance, leading to later functional intestinal disorders [7]. Aberrations of gut motility are often present in FGIDs, and currently an altered brain-gut axis has been accepted as a main pathogenetic mechanism of FGIDs, associated with dysfunction of the gastrointestinal autonomic nervous system [8]. Recently, an additional potential mechanism in the development of FGIDs seems to include alteration of the gut microbiota and its action on the modulation of intestinal inflammation, and the immune system activation [9, 10].

The Gastrointestinal Ecologic Unit

The gastrointestinal tract and organisms within its lumen constitute an ecologic unit. All of the components of the ecologic unit are important and depend on each other. In order to understand the mechanisms by which probiotics are involved in the maintenance of intestinal microbial microecology and, as consequence, are useful in FGIDs, the concepts of the human microbiome, dynamic intestinal barrier function, and neuroimmunology system have to be introduced.

The Human Microbiome
The human microbiome, broadly defined, is the full collection of microbes (bacteria, fungi, viruses, etc.) and their DNA that naturally exist in a given habitat of the human body. Several lines of evidence indicate that bacteria may be involved in the pathogenesis and pathophysiology of FGIDs through different actions (fig. 1; see below).

The Intestinal Barrier Function
The intact intestinal epithelium with the normal intestinal microflora represents a stable barrier for protecting the host and providing normal intestinal function. In fact, the gut barrier is able to prevent luminal pathogens and harmful substances from entering into the internal milieu, but also promotes digestion and different architectural units of this barrier [11]. The intestinal mucosa is constantly sampling luminal contents and making molecular adjustments at its frontier. This system also protects against pathogens by elaborating and releasing protective peptides, cytokines, chemokines, and phagocytic cells. When either the normal microflora or the epithelial cells are disturbed by triggers such as dietary antigens, pathogens, chemicals, or radiation, defects in the barrier mechanisms become evident. Altered permeability further facilitates the invasion of pathogens, foreign antigens, and other harmful substances [12].

Gastrointestinal Mucosal Immunology
Gut-associated lymphoid tissue is composed of both inductive (Peyer's patches) and effector sites (intraepithelial cells and lamina propria). Gut-associated lymphoid tissue, dealing with intestinal microflora, prevents potentially harmful intestinal antigens from reaching the systemic circulation and induces systemic tolerance against luminal antigens by a process that involves polymeric immunoglobulin A secretion and the induction of regulatory T cells. Recently, we began to appreciate that immunoglobulin A plays a key role in the regulation of bacterial communities in the intestine and that the repertoire of gut microbiota is closely linked to the proper functioning of the immune system. Antigen-presenting cells (macrophages, dendritic cells, B cells) efficiently take up and transport a variety of microorganisms and present antigen. Recognition of antigens by antigen-presenting cells triggers a family pattern of recognition receptors (Toll-like receptors) which change cell phenotype and function.

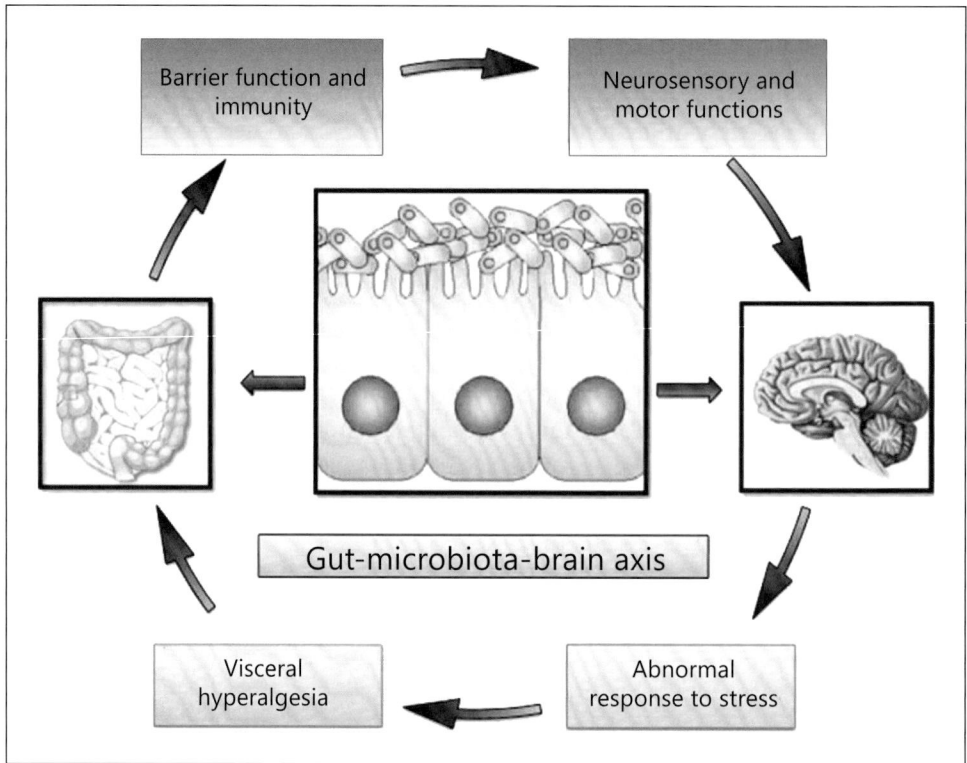

Fig. 1. Changes in the interaction between intestinal microbiota and host factors could be important for the pathophysiology of FGIDs. These factors in turn could be related to changes in homeostatic pathways including barrier function, neuromotor sensory function, and the gut-brain axis.

Toll-like receptors direct immune responses by activating signaling events leading to an elevated expression of cytokines and chemokines that recruit and regulate the immune and inflammatory cells, which in turn either initiate or enhance host immune responses [13].

Several studies have demonstrated a low grade of activation of innate and adaptive immune response in children with FGIDs [14, 15]. Moreover, the mucosal immune activation mediated by the microbiota has an implication in sensorimotor dysfunction of patients with FGIDs [16, 17].

The Enteric Nervous System and Gastrointestinal Motility
Brain-intestinal interactions are well-known mechanisms for the regulation of intestinal function in both healthy and diseased states. The brain can influence commensal organisms via changes in gastrointestinal motility, secretion, and intestinal permeability, or directly via signaling molecules released into the gut lumen from cells in the lamina propria (enterochromaffin cells, neurons, immune cells). Enteric microbiota communication occurs via epithelial-cell, receptor-mediated signaling and, when in-

testinal permeability is increased, through direct stimulation of host cells in the lamina propria. Enterochromaffin cells are integral to these communications and serve as bidirectional transducers that regulate communication between the intestinal lumen and the nervous system [18]. Disruption of the bidirectional interactions between the enteric microbiota and the nervous system may be involved in the pathophysiology of FGIDs.

Normal intestinal motility requires the coordination between the extrinsic neurons, enteric motor neurons, interstitial cells of Cajal, and smooth muscle cells. The enteric nervous system is a complex integrative brain (also called the second brain), which is capable of controlling gastrointestinal function. The enteric nervous system influences the gut by directly acting on contraction and indirectly influencing the cells of the gut immune system and the epithelial cells. This interaction is bidirectional and relies on the mechanisms of neuroimmune interaction, which involves the bacterial component of the activation of Toll-like as well as other bacterial molecular pattern receptors to trigger innate immune responses and the intestinal neural pathways [19].

Role of Probiotics in Functional Gastrointestinal Disorders

An aberrant gut microbial composition, such as low lactobacilli load and an increased concentration of coliforms, may play an important role in the pathogenesis of gastrointestinal stress-related disorders such as FGIDs. Interestingly, a recent study demonstrated that fecal microbiota in patients with irritable bowel syndrome (IBS) could be grouped in a cluster which was substantially different from that of healthy controls [20].

The postulated mechanisms of action of probiotics on the different mechanisms of pathophysiology of FGIDs include: improvement of the barrier function of the epithelium, suppression of the growth and binding of pathogenic bacteria, a positive effect on visceral hypersensitivity, and immunomodulatory effects [21]. Although the exact mechanisms are not well understood, an increasing body of evidence supports the role of probiotics in reducing anxiety and stress response, with some probiotics having the potential to decrease inflammatory cytokine concentration and decrease oxidative stress, and others altering the expression of the GABA receptor in the central nervous system [22]. Interestingly, vagotomy prevented these latter effects of probiotics, suggesting that parasympathetic innervation plays an important role in the transmission of information from gut to brain and also in homeostasis associated with the immune system.

It is now known that specific microbes alter pain pathways in the intestine, inducing the expression of opioid and cannabinoid receptors in intestinal epithelial cells and mediating analgesic functions in the gut, similar to the effects of morphine [23]. A role for the gut microbiota in pain perception has also been proposed in decreasing

Table 2. Probiotics in children with FGIDs

Study	Agent	Design	Sample size	Symptoms
Bausserman [25]	*Lactobacillus* GG	DBPC	50	↓ abdominal distension
Gawrońska [26]	*Lactobacillus* GG	DBPC	104	↓ pain frequency
Romano [27]	*Lactobacillus reuteri*	DBPC	60	↓ pain intensity
Guandalini [28]	VSL#3	DBPC-CO	64	↓ global score
Francavilla [29]	*Lactobacillus* GG	DBPC	141	↓ pain intestinal permeability

DBPC = Double-blind placebo-controlled; DBPC-CO = double-blind placebo-controlled crossover.

visceral hypersensitivity through modulation of dorsal root ganglion single unit activity to colon-rectal distension [24]. The most important randomized controlled trials in treatment of major FGIDs in children are summarized in table 2.

Different single strains and combinations of probiotic have been tested on children, resulting in different effects on different clinical manifestations of IBS. The studies reported were well conducted and probiotics appear to be efficacious in IBS, but the magnitude of benefit and the most effective species and strain are uncertain. The optimal dose, role of combination therapy, strain-specific activity, stability within the gastrointestinal tract, possible development of antibiotic resistance, and duration of therapy still need to be clarified. The inconsistencies between the studies underline the need to look at each probiotic product separately for specific conditions, symptoms, and patient populations.

The use of probiotics in children with IBS is more limited compared to the use in adults, but also suggestive of beneficial effects. In the pediatric age, the bacterial composition changes dramatically with any change in food habit and environmental factor. Moreover, the brain development does not cease at birth. During infancy the brain establishes the myriad synaptic connections that provide the essential substrate for functional brain networks that underlie perception, cognition, and action. A recent study [30] surprisingly reveals that the bacterial content of the gut, which changes rapidly after birth as food ingestion begins, can modulate brain developmental pathways. Thus, manipulating the microbiota in children could have an amplified effect on intestinal function acting on gut-brain axis that is still in a critical developmental window.

Conclusion

The effect of the intestinal microflora on the pathophysiology of FGIDs has become more evident over the last few years even though the exact mechanisms of interaction between the intestinal bacteria and host are still unknown. Probiotics might play important roles in maintaining gut homeostasis by modulating intestinal barrier function, immunity, and motility, and by influencing the gut-brain interaction. The role of intestinal microbiota in the pathogenesis of FGIDS could represent a promising field of research in the near future.

References

1 Drossman DA: The functional gastrointestinal disorders and the Rome III process. Gastroenterology 2006;130:1377–1390.

2 Hyman PE, Milla PJ, Benninga MA, Davidson GP, Fleisher DF, Taminiau J: Childhood functional gastrointestinal disorders: neonate/toddler. Gastroenterology 2006;130:1519–1526.

3 Rasquin A, Di Lorenzo C, Forbes D, et al: Childhood functional gastrointestinal disorders: child/adolescent. Gastroenterology 2006;130:1527–1537.

4 Jones MP, Dilley JB, Drossman D, Crowell MD: Brain-gut connections in functional GI disorders: anatomic and physiologic relationships. Neurogastroent Motil 2006;18:91–103.

5 Drossman DA, Whitehead WE, Toner BB, Diamant NE, Hu YJB, Bangdiwala SI, et al: What determines severity among patients with painful functional bowel disorders? Am J Gastroenterol 2000;95:974–980.

6 Levy RL, Whitehead WE, Walker LS, Von KM, Feld AD, Garner M, et al: Increased somatic complaints and health-care utilization in children: effects of parent IBS status and parent response to gastrointestinal symptoms 2. Am J Gastroenterol 2004;99:2442–2451.

7 Budavari AI, Olden KW: Psychosocial aspects of functional gastrointestinal disorders. Gastroenterol Clin North Am 2003;32:477–506.

8 Tack J, Talley NJ, Camilleri M, Holtmann G, Hu P, Malagelada JR, Stanghellini V: Functional gastroduodenal disorders. Gastroenterology 2006;130:1466–1479.

9 O'Hara AM, Shanahan F: The gut flora as a forgotten organ. EMBO Rep 2006;7:688–693.

10 Quigley EM: Therapies aimed at the gut microbiota and inflammation: antibiotics, prebiotics, probiotics, synbiotics, anti-inflammatory therapies. Gastroenterol Clin North Am 2011;40:207–222.

11 Lu L, Walker WA: Pathologic and physiologic interactions of bacteria with the gastrointestinal epithelium. Am J Clin Nutr 2001;73:1124S–1130S.

12 Cryan JF, O'Mahony SM: The microbiome-gut-brain axis: from bowel to behavior. Neurogastroenterol Motil 2011;23:187–192.

13 Mason KL, Huffnagle GB, Noverr MC, Kao JY: Overview of gut immunology. Adv Exp Med Biol 2008;635:1–14.

14 Ohman L, Simren M: Pathogenesis of IBS: role of inflammation, immunity and neuroimmune interactions. Nat Rev Gastroenterol Hepatol 2010;7:163–173

15 Barbara G, Cremon C, Carini G, et al: The immune system in irritable bowel syndrome. J Neurogastroenterol Motil 2011;17:349–359.

16 Barbara G, Wang B, Stanghellini V, et al: Mast cell-dependent excitation of visceral-nociceptive sensory neurons in irritable bowel syndrome. Gastroenterology 2007;132:26–37.

17 Cenac N, Andrews CN, Holzhausen M, et al: Role for protease activity in visceral pain in irritable bowel syndrome. J Clin Invest 2007;117:636–647.

18 Grenham S, Clarke G, Cryan JF, Dinan TG: Brain-gut-microbe communication in health and disease. Front Physiol 2011;2:94.

19 Rhee SH, Pothoulakis C, Mayer EA: Principles and clinical implications of the brain-gut-enteric microbiota axis. Nat Rev Gastroenterol Hepatol 2009;6:306–314.

20 Rajilic-Stojanovic M, Biagi E, Heilig HG, et al: Global and deep molecular analysis of microbiota signatures in fecal samples from patients with irritable bowel syndrome. Gastroenterology 2011;141:1792–1801.

21 Indrio F, Riezzo G, Raimondi F, Bisceglia M, Cavallo L, Francavilla R: The effects of probiotics on feeding tolerance, bowel habits, and gastrointestinal motility in preterm newborns. J Pediatr 2008;152:801–806.

22 Bravo JA, Forsythe P, Chew MV, Escaravage E, Savignac HM, Dinan TG, Bienenstock J, Cryan JF: Ingestion of Lactobacillus strain regulates emotional behavior and central GABA receptor expression in a mouse via the vagus nerve. Proc Natl Acad Sci USA 2011;108:16050–16055.

23 Rousseaux C, Thuru X, Gelot A, Barnich N, Neut C, Dubuquoy L, Dubuquoy C, Merour E, Geboes K, Chamaillard M, Ouwehand A, Leyer G, Carcano D, Colombel JF, Ardid D, Desreumaux P: *Lactobacillus acidophilus* modulates intestinal pain and induces opioid and cannabinoid receptors. Nat Med 2007;13:35–37.

24 Verdú EF, Bercik P, Verma-Gandhu M, Huang XX, Blennerhassett P, Jackson W, Mao Y, Wang L, Rochat F, Collins SM: Specific probiotic therapy attenuates antibiotic induced visceral hypersensitivity in mice. Gut 2006;55:182–190.

25 Bausserman M, Michail S: The use of Lactobacillus GG in irritable bowel syndrome in children: a double-blind randomized control trial. J Pediatr 2005; 147:197–201.

26 Gawrońska A, Dziechciarz P, Horvath A, Szajewska H: A randomized double-blind placebo-controlled trial of Lactobacillus GG for abdominal pain disorders in children. Aliment Pharmacol Ther 2007; 25: 177–184.

27 Romano C, Ferrau' V, Cavataio F, Iacono G, Spina M, Lionetti E, Comisi F, Famiani A, Comito D: *Lactobacillus reuteri* in children with functional abdominal pain (FAP). J Paediatr Child Health 2010, E-pub ahead of print.

28 Guandalini S, Magazzù G, Chiaro A, La Balestra V, Di Nardo G, Gopalan S, Sibal A, Romano C, Canani RB, Lionetti P, Setty M: VSL#3 improves symptoms in children with irritable bowel syndrome: a multicenter, randomized, placebo-controlled, double-blind, crossover study. J Pediatr Gastroenterol Nutr 2010;5:24–30.

29 Francavilla R, Miniello V, Magistà AM, De Canio A, Bucci N, Gagliardi F, Lionetti E, Castellaneta S, Polimeno L, Peccarisi L, Indrio F, Cavallo L: A randomized controlled trial of Lactobacillus GG in children with functional abdominal pain. Pediatrics 2010;126: 1445–1525.

30 Heijtz RD, Wang S, Anuar F, Qian Y, et al: Normal gut microbiota modulates brain development and behavior. Proc Natl Acad Sci USA 2011;108:3047–3052.

Flavia Indrio
Department of Pediatrics, University of Bari
Piazza Giulio Cesare
IT–70124 Bari (Italy)
E-Mail f.indrio@alice.it

Probiotics in the Prevention and Treatment of Diseases in Adults and Children

Guarino A, Quigley EMM, Walker WA (eds): Probiotic Bacteria and Their Effect on Human Health and Well-Being.
World Rev Nutr Diet. Basel, Karger, 2013, vol 107, pp 87–94 (DOI: 10.1159/000347200)

Functional Gastrointestinal Disorders in Adults

Eamonn M.M. Quigley

Alimentary Pharmabiotic Centre, University College Cork, Cork, Ireland

Abstract

Functional gastrointestinal disorders are common worldwide, are of unknown etiology, and, though non-fatal, may result in considerable impairment in quality of life. For some, a plausible hypothesis which suggests a role for the gut microbiota in their pathogenesis has been advanced. Examples include the role of *Helicobacter pylori* in functional dyspepsia (FD), the phenomenon of post-infectious irritable bowel syndrome (IBS), and descriptions of qualitative and quantitative alterations in the microbiota in IBS and chronic constipation. While probiotics have been used on an empiric basis by sufferers for the relief of a variety of functional symptoms for decades, it is only recently that large-scale clinical trials have been performed. While positive trends have emerged in a number of areas (alleviation of eradication therapy-related adverse events in FD, overall symptom relief, and reductions in bloating and flatulence in IBS), more large, high-quality, long-duration studies are required in most areas.

Copyright © 2013 S. Karger AG, Basel

The functional gastrointestinal disorders (FGIDs) comprise a heterogeneous group of symptom-based conditions that cause troubling symptoms which cannot be explained on an 'organic' basis. FGIDs tend to run relapsing and remitting courses over long periods of time and are usually classified according to their predominant presenting symptom(s). For the purposes of clinical research, if not clinical practice, this is usually performed according to the Rome Diagnostic Criteria [1]. An FGID may affect any part of the gastrointestinal tract, with irritable bowel syndrome (IBS), functional dyspepsia (FD), and chronic constipation being the most common symptom clusters/syndromes encountered world-wide.

The etiology of the FGIDs is, for the most part, unknown and presently there is no verified, unifying hypothesis to explain all FGIDs. By definition, FGIDs are not associated with any identifiable structural, morphological, or biochemical abnormality; there are, therefore, no biomarkers or imaging findings that can assist in their positive identification [2, 3].

Epidemiological studies have revealed the prevalence and potential impact of FGIDs. For example, the prevalence of IBS, the best characterized of the FGIDs, in the Western world ranges from 3 to 15% [4]. FGIDs can be very burdensome, affecting quality of life due to family disruption, impaired work performance, and psychological comorbidity. FGIDs can lead to increased health-care seeking behaviors with associated high costs resulting from frequent visits to primary care physicians, emergency departments, and specialist clinics as well as the performance of unnecessary tests and even surgery [5–7]. Another feature shared by all of the FGIDs is an absence of disease-modifying therapies; therefore, treatment is based primarily on attempting, usually with modest success, to relieve individual symptoms.

While probiotics have been used on an empirical basis for the alleviation of symptoms that would now be encompassed within a more formally defined FGID, evidence from clinical trials is largely confined to three areas: FD, IBS, and chronic constipation. Probiotics have also been advocated for the relief of individual symptoms such as bloating and flatulence.

Functional Dyspepsia

Among the various FGIDs, FD (previously termed nonulcer dyspepsia) has proven the most difficult to define. To the lay person, the term indigestion best encapsulates the concept of dyspepsia; arriving at a more precise terminology has proven a real challenge. Indeed, a variety of symptoms, ranging from heartburn to epigastric pain and including a variety of postprandial ills may be encompassed within this term. The subtype FD refers to those instances where no organic explanation is uncovered, most commonly by upper gastrointestinal endoscopy. Despite attempts by numerous committees and consensus groups to agree on a uniform clinical definition of FD, clinical trials continue to use different diagnostic terminology, rendering the interpretation of data challenging. The Rome II committee defined FD as the presence of abdominal pain or discomfort centered in the epigastrium, which cannot be explained by upper gastrointestinal investigation, and present for at least 12 weeks over the last 12 months [8].The more recent Rome III definition [9] required symptoms to be present for the last 3 months, with symptom onset at least 6 months before diagnosis. In a major shift in emphasis, it also proposed that FD comprised at least two distinct subgroups: the postprandial distress syndrome, which features postprandial fullness and early satiety, and the epigastric pain syndrome, which features a more constant and less meal-related pain syndrome [9]. In an attempt to separate FD from gastroesophageal reflux disease (GERD), patients with prominent heartburn were excluded from both Rome definitions of FD. Such a clear separation is often impossible on clinical grounds given the overlap that exists between these disorders; however, it is clearly evident that dyspeptic patients with predominant heartburn are those most likely to respond to acid suppression; thereby, supporting the value of identifying the predominant symptom in a given patient [10]. Of the various factors (e.g. acid, dysmotility, visceral hypersensitivity) that have been variably invoked in the pathophysiology of FD [11], *Helicobacter pylori* (HP) infection, whose role in FD remains controversial, is the one that might be amenable to a probiotic intervention. In relation to HP, two separate issues have been addressed: firstly, do probiotics promote the eradication of HP and, secondly, can probiotic therapy reduce or eliminate antibiotic-related adverse events and thus improve compliance with anti-HP therapeutic regimens? With regard to the former, there is some, but by no means consistent, evidence from both experimental models and clinical studies that probiotics can exert bactericidal activity against the organism and prevent its adherence to gastric mucus cells. Evidence for an associated clinical benefit is less consistent, but there is some evidence to suggest that coadministration of lactoferrin and/or a probiotic may enhance the efficacy of eradication therapy [12–17]. There is also some evidence that coadministration of a probiotic with eradication regimes, for HP, may reduce the incidence of related gastrointestinal side effects [18–20]. The very recent Maastricht IV/Florence Consensus Report concluded that 'certain probiotics and prebiotics show promising results as an adjuvant treatment in reducing side effects', but emphasized that additional studies were needed [20].

Irritable Bowel Syndrome

IBS is the most common and best characterized of the functional bowel disorders. It typically presents with abdominal pain or discomfort that is associated with defecation or a change in bowel habit, and is often accompanied by features of disordered defecation as well as bloating and/or distension. Approximately 10–20% of adults in the West have symptoms consistent with IBS [21, 22]. A combination of visceral hypersensitivity, smooth muscle spasm, and impairment of central pain processing [23, 24] likely contribute to the pain associated with IBS, while altered intestinal motility probably underlies the disordered defecation experienced by some patients with IBS [25]. More recently, the possibility that alterations in the microbiota or in host-microbiota interactions might play a fundamental role in IBS has begun to gain credence [26, 27] and has provided a scientific rationale for the use of probiotics and other interventions that modify the microbiota in IBS [28]. Indeed, and reflecting perhaps the paucity of truly disease-modifying therapies that are available to relieve the disorder, irritable bowel sufferers commonly have recourse to the use of complementary and alternative medical remedies and practices [29]. Foremost among such approaches have been various dietary manipulations, including exclusion diets, and a variety of dietary supplements. In Europe, in particular, where several such products are advertised widely for their general 'immune-boosting' and 'health-enhancing' properties, probiotics have been widely used as dietary supplements by IBS patients. Recently, based on data from the experimental laboratory, as well as some evidence from clinical trials, the concept of probiotic use in IBS has begun to wend its way into the realm of conventional medicine. While probiotics have been used on an empirical basis for some time in the management of IBS, several recent developments have provided a more logical basis for their use in this context [30]. These include the clear recognition that IBS may be induced by bacterial gastroenteritis (postinfectious IBS) and that qualitative changes in the microbiota, as well as immune dysfunction, may be prevalent in IBS, in general.

Up to the year 2000, a small number of studies evaluated the response of IBS to probiotic preparations and, while results between studies were difficult to compare because of differences in study design, probiotic dose and strain, there was some, but by no means consistent, evidence of symptom improvement [31].

Since then, further studies have assessed the response to a number of well-characterized organisms and have produced discernible trends. Thus, a number of organisms, such as *Lactobacillus* GG, *Lactobacillus plantarum*, *Lactobacillus acidophilus, Lactobacillus casei,* the probiotic 'cocktail' VSL#3, and *Bifidobacterium animalis*, have been shown to alleviate individual IBS symptoms such as bloating, flatulence, and constipation, but only a few products have been shown to affect pain and global symptoms in IBS [32–36]. Among these, *Bifidobacterium infantis* 35624 has attracted particular attention [36]. In the first study with this organism, superiority was shown over both a lactobacillus and placebo for each of the cardinal symptoms of IBS (abdominal pain/discomfort, distension/bloating, and difficult defecation), as well as for a composite score [37]. A larger, 4-week duration, dose-ranging study of the same bifidobacterium in over 360 community-based subjects with IBS confirmed efficacy for this organism in a dose of 10^8; again, all of the primary symptoms of IBS were significantly improved and a global assessment of IBS symptoms at the end of therapy revealed a greater than 20% therapeutic gain for the effective dose of the probiotic over the placebo [38]. Further large, long-term, randomized controlled trials of this bifidobacterium and other strains are warranted in IBS and detailed explorations of its mechanism(s) of action are indicated.

Several factors, including a reduction in gas production [39–41], changes in bile salt conjugation, an antibacterial or antiviral effect (in the case of postinfectious IBS), the promotion of motility [42], effects on mucus secretion, or even a specific anti-inflammatory effect [37], could be relevant to the benefits of specific probiotic strains in IBS.

Constipation

Data on the impact of probiotics on constipation is scanty, to say the least. However, the demonstration that certain strains of bifidobacteria can accelerate whole gut and colon transit [42, 43] provides a rationale for the use of probiotics in constipation, as does limited data suggesting the presence of dysbiosis in the colon in chronic constipation [44]. Very few double-blind placebo-controlled trials of probiotics in acute or chronic constipation per se (i.e. other than in association with IBS) are available. In one of the few such studies, Koebnick et al. [45] documented a positive benefit for a probiotic beverage containing *L. casei* Shirota in patients with chronic constipation. Other studies which were uncontrolled, or in which the probiotic was combined with some other form of therapy, reported variable benefits for bifidobacteria, lactobacilli, and proprionibacteria, and infusions of fecal suspensions [46–51]. Chmielewska and Szajewska [52] performed a systematic review of randomized controlled trials of probiotics in functional constipation and concluded, not surprisingly, that until more data become available, the use of probiotics for the treatment of constipation condition should be considered investigational.

Bloating and Flatulence

The pathogenetic mechanisms of bloating, on the one hand, and flatulence, on the other, in IBS or other functional disorders are quite distinct; flatulence resulting in large part from increased bacterial fermentation in the colon (i.e. increased gas production), whereas bloating and distension do not appear to be associated with an overall increase in gas production, but may rather reflect impaired gas transit through the colon, combined, perhaps, with a very localized boost in fermentation [53]. Agrawal et al. [42] examined in detail the impact of *Bifidobacterium lactis* DN-173 010 on a number of gastrointestinal symptoms, including bloating, and demonstrated parallel improvements in bloating and a reduction in abdominal distension, as measured objectively by plethysmography. Based, in part, on the mistaken assumption that an alteration of the colonic flora may alleviate 'gas-related' symptoms and on the finding of a positive impact on bloating in a prior study [54], IBS patients with prominent bloating were specifically selected for probiotic therapy with the cocktail VSL#3 by Kim et al. [55]. Interestingly, while bloating did not improve, flatulence was reduced. In both of our studies with *B. infantis* 35624, bloating was shown to improve [37, 38]. Based on our current understanding of the pathogenesis of bloating and of results with probiotic species and strains that are effective in IBS, there appears to be no rational basis for restricting probiotic therapy in IBS to those who bloat or experience abdominal distension.

References

1 Rome III Criteria. Rome III Diagnostic Criteria for Functional Gastrointestinal Disorders. http://www.romecriteria.org/assets/pdf/19_RomeIII_apA_885–898.pdf (accessed 10 January 2013).

2 Drossman DA: The Rome Foundation and Rome III. Neurogastroenterol Motil 2007;19:783–786.

3 Talley N, Spiller RC: Irritable bowel syndrome: a little understood organic bowel disease? Lancet 2002; 360:555–556.

4 Cremonini F, Talley NJ: Irritable bowel syndrome: epidemiology, natural history, health-care seeking and emerging risk factors. Gastroenterol Clin North America 2005;34:189–204.

5 Talley NJ: Functional gastrointestinal disorders as a public health problem Neurogastroenterol Motil 2008;20:(suppl 1):121–129.

6 Longstreth GF, Wilson A, Knight K, et al: Irritable bowel syndrome, health care use, and costs: a U.S. managed care perspective. Am J Gastroenterol 2003; 98:6007.

7 Talley N, Gabriel S, Harmsen W, Zinsmeister A, Evans R: Medical costs in community subjects with irritable bowel syndrome. Gastroenterology 1995;109: 1736–1741.

8 Talley NJ, Stanghellini V, Heading RC, Koch KL, Malagelada JR, Tytgat GN: Functional gastroduodenal disorders. Gut 1999;45(suppl 2):II37–II42.

9 Tack J, Talley NJ, Camilleri M, Holtmann G, Hu P, Malagelada JR, Stanghellini V: Functional gastroduodenal disorders. Gastroenterology 2006;130:1466–1479.

10 Quigley EM, Lacy BE: Overlap of functional dyspepsia and GERD-diagnostic and treatment implications. Nat Rev Gastroenterol Hepatol 2013, E-pub ahead of print.

11 Quigley EM, Keohane J: Dyspepsia. Curr Opin Gastroenterol 2008;24:692–697.

12 Gotteland M, Brunser O, Cruchet S: Systematic review: are probiotics useful in controlling gastric colonization by *Helicobacter pylori*? Aliment Pharmacol Ther 2006;23:1077–1086.

13 Zou J, Dong J, Yu X: Meta-analysis: Lactobacillus containing quadruple therapy versus standard triple first-line therapy for *Helicobacter pylori* eradication. Helicobacter 2009;14:97–107.

14 Sachdeva A, Nagpal J: Meta-analysis: efficacy of bovine lactoferrin in *Helicobacter pylori* eradication. Aliment Pharmacol Ther 2009;29:720–730.

15 Tong JL, Ran ZH, Shen J, et al: Meta-analysis: the effect of supplementation with probiotics on eradication rates and adverse events during *Helicobacter pylori* eradication therapy. Aliment Pharmacol Ther 2007;25:155–168.

16 Sachdeva A, Nagpal J: Effect of fermented milk-based probiotic preparations on *H. pylori* eradication: a systematic review and meta-analysis of randomized-controlled trials. Eur J Gastroenterol Hepatol 2009;1: 45–53.

17 Szajewska H, Horvath A, Piwowarczyk A: Meta-analysis: the effects of *Saccharomyces boulardii* supplementation on *Helicobacter pylori* eradication rates and side effects during treatment. Aliment Pharmacol Ther 2010;32:1069–1079.

18 Armuzzi A, Cremonini F, Bartolozzi F, et al: The effect of oral administration of Lactobacillus GC on antibiotic-associated gastrointestinal side-effects during *Helicobacter pylori* eradication therapy. Aliment Pharmacol Ther 2001;15:163–169.

19 Myllyluoma E, Veijola L, Ahlroos T, Tynkkynen S, Kankuri E, Vapaatalo H, Rautelin H, Korpela R: Probiotic supplementation improves tolerance to *Helicobacter pylori* eradication therapy – a placebo-controlled, double-blind randomized pilot study. Aliment Pharmacol Ther 2005;21:1263–1272.

20 Malfertheiner P, Megraud F, O'Morain CA, Atherton J, Axon AT, Bazzoli F, Gensini GF, Gisbert JP, Graham DY, Rokkas T, El-Omar EM, Kuipers EJ, European Helicobacter Study Group: Management of *Helicobacter pylori* infection – the Maastricht IV/ Florence Consensus Report. Gut 2012;61:646–664.

21 Saito YA, Schoenfeld P, Locke GR 3rd: The epidemiology of irritable bowel syndrome in North America: a systematic review. Am J Gastroenterol 2002;97: 1910–1915.

22 Quigley EM, Bytzer P, Jones R, Mearin F: Irritable bowel syndrome: the burden and unmet needs in Europe. Dig Liver Dis 2006;38:717–723.

23 Trimble KC, Farouk R, Pryde A, Douglas S, Heading RC: Heightened visceral sensation in functional gastrointestinal disease is not site-specific. Evidence for a generalized disorder of gut sensitivity. Dig Dis Sci 1995;40:1607–1613.

24 Aziz Q, Thompson DG, Ng VW, et al: Cortical processing of human somatic and visceral sensation. J Neurosci 2000;20:2657–2663.

25 McKee DP, Quigley EM: Intestinal motility in irritable bowel syndrome: is IBS a motility disorder? Part 2. Motility of the small bowel, esophagus, stomach, and gall-bladder. Dig Dis Sci 1993;38:1773–1782.

26 Aziz Q, Doré J, Emmanuel A, Guarner F, Quigley EM: Gut microbiota and gastrointestinal health: current concepts and future directions. Neurogastroenterol Motil 2013;25:4–15.

27 Quigley EM: Bugs on the brain; brain in the gut-seeking explanations for common gastrointestinal symptoms. Ir J Med Sci 2012;182:1–6.

28 Jeffery IB, Quigley EMM, Ohman L, Simrén M, O'Toole PW: The microbiota link to irritable bowel syndrome: an emerging story. Gut Microbes 2012;3: 572–576.

29 Hussein Z, Quigley EM: Complementary and alternative medicine in irritable bowel syndrome. Aliment Pharmacol Ther 2006;15:465–471.

30 Quigley EMM, Flourie B: Probiotics in irritable bowel syndrome: a rationale for their use and an assessment of the evidence to date. Neurogastroenterol Motil 2007;19:166–172.

31 Hamilton-Miller JMT: Probiotics in the management of irritable bowel syndrome: a review of clinical trials. Microb Ecol Health Dis 2001;13:212–216.

32 Nikfar S, Rahimi R, Rahimi F, Derakhshani S, Abdollahi M: Efficacy of probiotics in irritable bowel syndrome: a meta-analysis of randomized, controlled trials. Dis Colon Rectum 2008;51:1775–1780.

33 McFarland LV, Dublin S: Meta-analysis of probiotics for the treatment of irritable bowel syndrome. World J Gastroenterol 2008;14:2650–2661.

34 Hoveyda N, Heneghan C, Mahtani KR, Perera R, Roberts N, Glasziou P: A systematic review and meta-analysis: probiotics in the treatment of irritable bowel syndrome. BMC Gastroenterol 2009 Feb 16;9: 15.

35 Moayyedi P, Ford AC, Talley NJ, Cremonini F, Foxx-Orenstein A, Brandt L, Quigley EMM: The efficacy of probiotics in the therapy of irritable bowel syndrome: a systematic review. Gut 2010;59:325–332.

36 Brenner DM, Moeller MJ, Chey WD, Schoenfeld PS: The utility of probiotics in the treatment of irritable bowel syndrome: a systematic review. Am J Gastroenterol 2009;104:1033–1049.

37 O'Mahony L, McCarthy J, Kelly P, et al: A randomized, placebo-controlled, double-blind comparison of the probiotic bacteria lactobacillus and bifidobacterium in irritable bowel syndrome (IBS): symptom responses and relationship to cytokine profiles. Gastroenterology 2005;128:541–551.

38 Whorwell PJ, Altringer L, Morel J, Bond Y, Charbonneau D, O'Mahony L, Kiely B, Shanahan B, Quigley EM: Efficacy of an encapsulated probiotic Bifidobacterium infantis 35624 in women with irritable bowel syndrome. Am J Gastroenterol 2006;101: 326–333.

39 Biarti B, Matterelli P: The family Bifidobacteria; in Dworkin M, Fallow S, Rosenberg E, Schieifer KH, Stackebrandt E (eds): The Prokaryotes. New York, Springer, 2001, pp 1–70.

40 Sen S, Mulan MM, Parker TJ, et al: Effect of Lactobacillus plantarum 299V on colonic fermentation and symptoms of irritable bowel syndrome. Dig Dis Sci 2002;47:2615–2620.

41 Barrett JS, Canale KE, Gearry RB, Irving PM, Gibson PR: Probiotic effects on intestinal fermentation patterns in patients with irritable bowel syndrome. World J Gastroenterol 2008;14:5020–5024.

42 Agrawal A, Houghton LA, Morris J, Reilly B, Guyonnet D, Goupil Feuillerat N, Schlumberger A, Jakob S, Whorwell PJ: Clinical trial: the effects of a fermented milk product containing Bifidobacterium lactis DN-173 010 on abdominal distension and gastrointestinal transit in irritable bowel syndrome with constipation. Aliment Pharmacol Ther 2009; 29:104–114.

43 Waller PA, Gopal PK, Leyer GJ, Ouwehand AC, Reifer C, Stewart ME, Miller LE: Dose-response effect of Bifidobacterium lactis HN019 on whole gut transit time and functional gastrointestinal symptoms in adults. Scand J Gastroenterol 2011;46:1057–1064.

44 Khalif IL, Konovitch EA, Maximova ID, Quigley EMM: Alterations in the colonic flora and intestinal permeability and evidence of immune activation in chronic constipation. Dig Liver Dis 2005;37:838–849.

45 Koebnick C, Wagner I, Leitzmann P, Stern U, Zunft HJ: Probiotic beverage containing Lactobacillus casei Shirota improves gastrointestinal symptoms in patients with chronic constipation. Can J Gastroenterol 2003;17:655–659.

46 Riezzo G, Orlando A, D'Attoma B, Guerra V, Valerio F, Lavermicocca P, De Candia S, Russo F: Randomised clinical trial: efficacy of Lactobacillus paracasei-enriched artichokes in the treatment of patients with functional constipation – a double-blind, controlled, crossover study. Aliment Pharmacol Ther 2012;35:441–450.

47 Amenta M, Cascio MT, Di Fiore P, Venturini I: Diet and chronic constipation. Benefits of oral supplementation with symbiotic zir fos (Bifidobacterium longum W11 + FOS Actilight). Acta Biomed 2006;77: 157–162.

48 Banaszkiewicz A, Szajewska H: Ineffectiveness of Lactobacillus GG as an adjunct to lactulose for the treatment of constipation in children: a double-blind, placebo-controlled randomized trial. J Pediatr 2005;146:364–369.

49 Ouwehand AC, Lagstrom H, Suomalainen T, Salminen S: Effect of probiotics on constipation, fecal azoreductase activity and fecal mucin content in the elderly. Ann Nutr Metab 2002;46:159–162.

50 Borody TJ, Warren EF, Leis SM, Surace R, Ashman O, Siarakas S: Bacteriotherapy using fecal flora: toying with human motions. J Clin Gastroenterol 2004; 38:475–483.

51 Choi SC, Kim BJ, Rhee PL, Chang DK, Son HJ, Kim JJ, Rhee JC, Kim SI, Han YS, Sim KH, Park SN: Probiotic fermented milk containing dietary fiber has additive effects in IBS with constipation compared to plain probiotic fermented milk. Gut Liver 2011;5: 22–28.

52 Chmielewska A, Szajewska H: Systematic review of randomised controlled trials: probiotics for functional constipation. World J Gastroenterol 2010;16: 69–75.

53 Quigley EM: Germs, gas and the gut; the evolving role of the enteric flora in IBS. Am J Gastroenterol 2006;101:334–335.

54 Kim HJ, Camilleri M, McKenzie S, Lempke MB, Burton DD: A randomized controlled trail of a probiotic, VSL#3 on gut transit and symptoms in diarrhoea-predominant IBS. Aliment Pharmacol Ther 2003;17: 895–904.

55 Kim HJ, Vazquez Roque MI, Camilleri M, Stephens D, Burton DD, Baxter K, Thomforde G, Zinsmeister AR: A randomised controlled trial of probiotic combination VSL#3 and placebo in IBS with bloating. Neurogastroenterol Motil 2005;17:687–696.

Eamonn M.M. Quigley, MD, FRCP, FACP, FACG, FRCPI
Alimentary Pharmabiotic Centre
University College Cork
Cork (Ireland)
E-Mail e.quigley@ucc.ie

Probiotics in the Prevention and Treatment of Diseases in Adults and Children

Guarino A, Quigley EMM, Walker WA (eds): Probiotic Bacteria and Their Effect on Human Health and Well-Being.
World Rev Nutr Diet. Basel, Karger, 2013, vol 107, pp 95–102 (DOI: 10.1159/000345738)

Metabolic Syndrome and Obesity in Children

R. Luoto[a, b] · M.C. Collado[f] · K. Laitinen[c, d] · S. Salminen[d] ·
E. Isolauri[b, e]

[a] Department of Paediatrics, Satakunta Central Hospital, Pori, [b]Institute of Clinical Medicine,
[c]Institute of Biomedicine, [d]Functional Foods Forum, University of Turku, and [e]Department of Paediatrics,
Turku University Hospital, Turku, Finland; [f]Instituto de Agroquímica y Tecnología de Alimentos (IATA-CSIC),
Valencia, Spain

Abstract

Overweight and obesity can be currently considered a major threat to human well-being. Especially the increasing incidence of obesity in children adumbrates its continuance into adulthood and thus escalation of the problem in the future. Recent scientific advances have highlighted one link between food intake and weight gain: specific strains of the gut microbiota increase energy efficiency by affecting the harvesting and storage of energy from the diet and, furthermore, set the inflammation tone to hamper insulin sensitivity, thereby favoring a metabolic syndrome associated with obesity. During the first few days of life, the microbial colonization of the human intestine is a rapid and varying process. In view of the high immunoreactive capacity of a constant and massive luminal antigen challenge, microbial contact during a critical phase of development may deviate towards inflammatory response instead of tolerance, and manifest itself in allergic, autoimmune, and chronic inflammatory diseases, such as obesity. Hence, gut microbiota modulation with probiotics is currently attracting interest in the search for preventive and therapeutic applications in weight management.

Copyright © 2013 S. Karger AG, Basel

Obesity constitutes a formidable pediatric health problem; in fact, it is the most widely prevalent nutritional disorder among children throughout the world. Notwithstanding the extensive and multidisciplinary scientific interest focusing on this problem, research has so far been unable to conclusively identify the determinants underlying this epidemic. Rather, an escalation of the disorder is to be expected since the velocity of propagation is highest in the pediatric population.

The root cause of obesity is energy imbalance: more calories are consumed than expended. This conception of a direct linkage is, however, challenged by evidence accumulating from epidemiological studies: the foundation of obesity is already laid during the fetal period or the first months of life, the nutritional environment possibly

Fig. 1. The gut microbiota programs host defense and plays an important role in human health by impacting nutrient utilization, enhancing gut barrier function, and stimulating immune development.

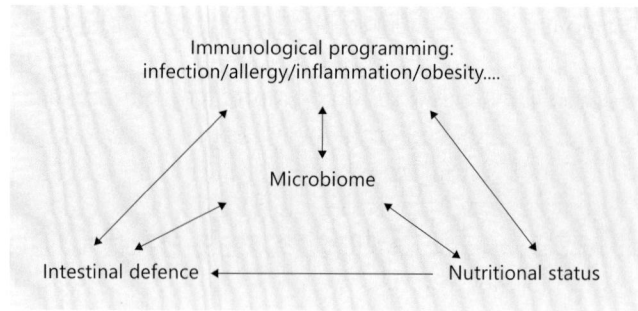

permanently altering the body's structure, physiology, and metabolism, thus leading to disease in adult life [1]. In addition to nutritional programming, this developmental concept also involves hormonal, metabolic, immunological, and microbial programming. The mother may transfer environmental information to the fetus through the placenta or to her infant through lactation. This shaping provenance may include the mother's unbalanced dietary intake, weight status (under- or overnutrition), or microbiota composition.

The gut microbiota has recently come to be seen as a conductor functioning at the intersection between host genotype and diet to modulate the host physiology impacting nutrient utilization, absorption, and metabolism (fig. 1). The compositional development of the gut microbiota, a substantial antigen challenge for the newborn, is essential for the developmental regulation of the intestinal physiology, especially of the epithelial barrier, and for the maturation of the gut-associated lymphoid tissue [2]. Exposure to microbes during pregnancy and during the first months of life may influence the metabolic and immunologic profile of the offspring and consequently the risk of chronic inflammatory diseases such as obesity developing later in life. In this context, administration of probiotics during the perinatal period and lactation with a view to modulating the intestinal microbiota composition may provide a novel tool in addressing the obesity epidemic.

Compositional Development of the Gut Microbiota: Pre-, Peri- and Postnatal Determinants

The general conception has long been that the gut microbiota starts to establish itself after birth, converges toward an adult-like microbiota by the end of the first year of life, and fully resembles the adult microbiota by 2 years of age; this step-wise process being modified by diet and environment during infancy [3]. Initial colonizers are generally facultative anaerobes including enterobacteria, coliforms, lactobacilli, and streptococci, which are then replaced by anaerobic genera such as *Bifidobacterium*, *Bacteroides*, *Clostridium*, and *Eubacterium* by the end of the first week of life [3].

Luoto · Collado · Laitinen · Salminen · Isolauri

The maternal microbiota forms the first inoculum to the development of the child's microbiota via direct contact with maternal vaginal and gastrointestinal microbiota. Vaginally delivered infants acquire bacterial communities resembling their own mother's vaginal microbiota, whereas cesarean section infants acquire bacterial communities similar to those found on the skin.

Other factors may also exert an influence on the compositional development of the gut microbiota, such as the environment and gestational age at birth, the use of antibiotics early in life, and the mode of feeding [3]. Breast milk has been shown to be an excellent and continuous source of commensal and potentially beneficial bacteria, including staphylococci, streptococci, lactic acid bacteria, and bifido-bacteria, while other components such as oligosaccharides and unique proteins support the growth of specific bacteria, most importantly *Bifidobacteria*, which typify the gut microbiota of the healthy breast-fed infant [4]. Moreover, bioactive molecules in breast milk contribute to microbial recognition with innate immune receptors such as Toll-like receptors, which are key mediators in the communication between the host and colonizing microbes, leading to a tolerogenic immune milieu.

Counter to this general dogma, microbial contact of the human body may in fact begin already before birth. Evidence has been reported that indicates traces of microbes, i.e. microbial DNA from intestinal bacteria, are detectable in the placenta, amniotic fluid, and fetal membranes in term pregnancies without signs of inflammation, rupture of membranes, or onset of labor [5]. Furthermore, microbial DNA has also been found in the meconium of healthy term neonates, suggesting a prenatal origin [6]. The immunomodulatory properties of maternal microbial compounds can further modulate the innate immune functions in the fetoplacental unit. Maternal microbial contact during pregnancy has been linked experimentally to the activation status of the T lymphoid cells present in the gut at birth [7] and also clinically to alterations in Toll-like receptor-related innate immune gene expression in the fetal gut [8]. Interestingly, not only the mother's intestinal microbiota, but also body weight, weight gain during pregnancy, and metabolic biomarkers have been shown to be associated with the initial microbial inoculum for the newborn and, furthermore, with the breast milk composition, thus affecting the early immune responses of the naive immune system [9, 10].

Alteration in the compositional development of the gut microbiota of a newborn has been clinically demonstrated in a few studies to predispose to diseases later in life. Aberrancies in the early microbiota have already been shown to be associated with a higher risk of allergy and gut inflammatory conditions, and recently also with obesity [11–13]. Furthermore, children born by cesarean section have been shown to have an increased risk of chronic inflammatory conditions such as celiac disease, type 1 diabetes mellitus, and asthma as compared with children born by vaginal delivery [14].

Evidence of Gut Microbiota Influences on Obesity in Infants and Children

Although genetic factors can determine the propensity of an individual to become obese, the recent increase in the phenomenon probably reflects environmental and lifestyle changes, with dietary habits being a major contributor. Altered dietary intake not only affects energy balance, but also has a major impact on the gut microbial composition, which can play an instrumental role in the control of host body weight and energy metabolism [15], and furthermore favor the onset of a low-grade inflammatory state and metabolic diseases, especially among individuals predisposed to obesity [16]. Pioneer experimental studies providing novel evidence of metabolic activities of the gut microbiota have demonstrated firstly that the gut microbiota facilitates the extraction of calories from the ingested diet and their storage in the host adipose tissue for later use, and secondly that characteristic alterations in the gut microbiota composition take place in obese compared with lean mice [15, 17, 18]. Thirdly, the transferable nature of the obese phenotype by transplantation of obese flora to germ-free mice has been recognized [19].

Recently, a relative abundance or paucity of various types of gut bacteria in obese and lean humans has also been demonstrated by a number of groups [reviewed in detail in 20]. In fact, a connection between a relative increase in the abundance of bacteria belonging to the phylum Firmicutes and a reduction in the level of Bacteroidetes with obesity has been proposed, but in light of the most recent findings, the possibility has been envisaged that smaller changes in the gut microbiota community, rather than those occurring at wide phylum levels, might be involved in overweight development. Additionally, dietary changes have been shown to modify gut microbiota composition and activity [20]. It has also been suggested that an aberrant gut microbiota composition is involved in the initiation of a low-grade inflammatory state and consequently in obesity-associated metabolic disorders [16]. As a corollary, an increase in *Bifidobacterium* spp. was demonstrated to correlate positively with normalization of the inflammatory status [16].

A gut microbiota profile favoring a higher number of bifidobacteria and a lower number of *Staphylococcus aureus* has been shown to be associated with protection against maternal overweight development during pregnancy, which itself is linked with changes in the gut microbiota. This tends to confirm a direct interrelationship between weight gain and microbiota changes [21, 22]. Additionally, the infant fecal microbial composition has been demonstrated to be related to maternal weight and weight gain over pregnancy [10]. In the study in question, Bacteroidetes and *Staphylococcus* levels were significantly higher in infants from overweight mothers during the first 6 months of life. Further, infants of women with excessive weight gain over pregnancy showed lower levels of bifidobacteria than those born to women with normal weight gain. A mother's higher weight and BMI were related in the study population to higher concentrations of Bacteroidetes, *Clostridium,* and *Staphylococcus,* and lower levels of the *Bifidobacterium* group [10]. The instrumental role of microbial stimulus during pregnancy to the later metabolic programming of the offspring may also partly explain the findings in a large prospective cohort study published by Lawlor et al. [23]. Most of the association between maternal weight gain during pregnancy and later offspring BMI proved attributable to intrauterine mechanisms other than shared familial (genetic and early environmental) characteristics in overweight and obese women.

These bacterial differences may culminate during the perinatal period, where translocation of commensal bacteria appears to be particularly vigorous [24]. Bacterial translocation may also be endorsed by labor and early lactation, related to hormonal or stress-induced adaptation. Lack of these physiological adaptations may partly explain the increased risk of obesity reported in four large Brazilian birth cohorts, which showed a reduction in subsequent obesity development in children who were delivered with vaginally compared with cesarean section [25, 26]. Moreover, higher levels of the *Staphylococcus* group and lower levels of the *Bifidobacterium* group have been detected in the breast milk of overweight compared to normal-weight mothers, indicating an additional mechanism explaining the heightened obesity risk in infants of overweight mothers [10]. Further, complex interactions of cytokines and microbiota in breast milk have been reported, as TGF-2 and sCD14 levels in the breast milk of overweight mothers have tended to be lower than in normal-weight mothers [10]. The prevalence of *Akkermansia muciniphila*-type bacteria was higher in overweight mothers, and their numbers were related to the IL-6 concentration in the colostrum, which was also related to lower counts of *Bifidobacterium* group bacteria in excessive-weight women [10].

On the other hand, recent molecular work has characterized the previously unknown microbial diversity of mature milk [27], although its full composition through time and the factors shaping it remain unknown. The milk microbial community has been characterized by 454 pyrosequencing in mothers who varied in BMI and weight gain over pregnancy [28]. Pyrosequencing of the 16S rRNA gene demonstrated a wide bacterial diversity already in the colostrum, dominated by lactic acid bacteria, with variations at 1 and 6 months of breastfeeding, which would indicate significant changes in the milk microbiome composition and diversity in a manner similar to what has been reported for other milk components such as protein and fat content [29]. Specific shifts in bacterial composition were associated with maternal BMI and weight gain, as confirmed by qPCR, showing that milk from obese mothers tended to contain a different and less diverse bacterial community compared to that of normal weight mothers.

The clinical course whereby the complex host-microbiota interaction guides the metabolic programming of the child was documented in a 7-year survey of children followed in one clinical study. Children who became overweight by 7 years of age had lower levels of bifidobacteria and higher levels of *S. aureus* at 6 and 12 months of age compared to those remaining normal weight [11]. In a subgroup of this same cohort, a trend towards lower numbers of fecal *Bifidobacteria* at the age of 3 months was shown to be associated with subsequent overweight development over 10 years of follow-up, with excessive weight gain beginning in these children immediately after 3 weeks of age [30].

The Role of Probiotics in Reshaping the Early Gut Microbiota

Modification of the gut microbiota by probiotics early in life has attracted scientific interest since there is a critical period during the first months of life which affords an important opportunity for immune education, when the establishment of the intestinal microbiota and maturation of the immune system are not yet completed. The administration of probiotics during pregnancy is also under consideration in view of the positive effects some strains exert on certain clinical conditions both in pregnant women and in the child. To date, however, the clinical data remain scant.

Taking into account that early microbial colonization can have far-reaching impacts on infant weight development. It is of note that a perinatal probiotic intervention with *Lactobacillus rhamnosus* GG (LGG) has been demonstrated to moderate excessive weight gain. This is especially so among children who later became overweight during the first years of life, with the impact being most pronounced at the age of 4 years [31]. In another placebo-controlled clinical study, a perinatally administered probiotic combination (LGG + *B. lactis* Bb 12) was demonstrated to attain consistently improved plasma glucose concentrations and insulin sensitivity in healthy women during pregnancy and 12 months postpartum when an advantageous dietary intake was combined with probiotics [32]. This same intervention study provided clinical evidence that probiotic consumption lowers the risk of maternal central adiposity over the 6-month postpartum period [33]. Furthermore, the beneficial effects were shown to extend to neonates and infants. Interestingly, nutrition counseling and probiotic intervention have been demonstrated to have a distinct effect on gestational diabetes, with probiotics reducing the risk and dietary counseling reducing the risk of fetal overgrowth associated with it [34].

Conclusion

Overall, it is acknowledged that perturbation of the gut microbiota composition, also known as dysbiosis, might be a pivotal factor and the 'missing link' in the fight against the obesity epidemic. However, before dysbiosis can be established, the composition of a healthy 'normal' microbiota has to be defined in evaluating the compositional development of the gut microbiota in healthy breast-fed infants who also remain healthy in the long term. Considering that the maternal microbiota forms the first inoculum for the development of the child's microbiota and that the gut microbiota composition may play a pivotal role in the microbial, metabolic, and immunological programming of the infant, the maternal as well as the infant's early intestinal microbiota may influence the subsequent development of overweight in children. On this basis, specific strategies to modify the gut microbiota with probiotics to enhance the bifidobacteria population may thus emerge as a measure to retard the incidence of overweight development, and as a corollary, restrain the Western lifestyle disease epidemic (fig. 2). The health-ame-

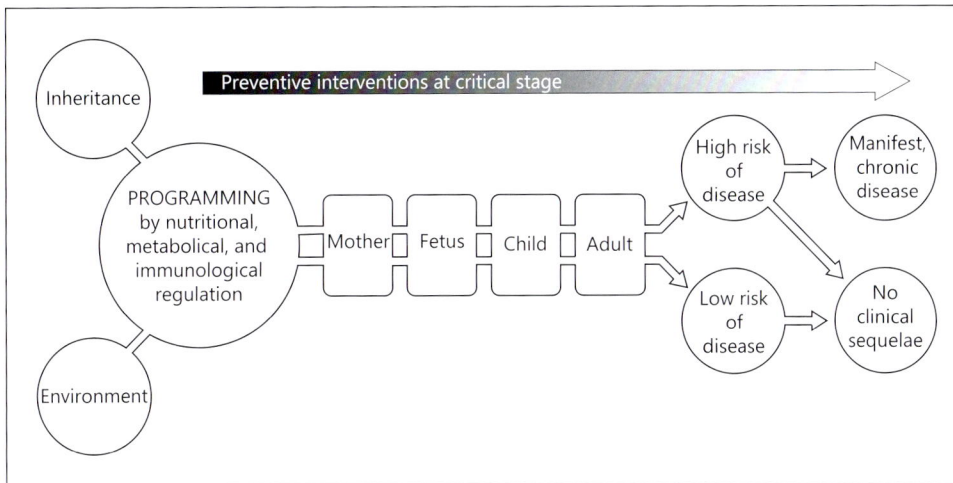

Fig. 2. Preventive interventions at the critical stage during pregnancy and breast-feeding might provide environmental influences resulting in the programming of a low risk of disease.

liorating effects of probiotics can be considered to be most pronounced during perinatal or early postnatal life, i.e. during the critical period when the composition of the gut microbiota and immunological responsiveness is being consolidated.

References

1 Barker DJ: The origins of the developmental origins theory. J Intern Med 2007;261:412–417.

2 Brandtzaeg P: Development and basic mechanisms of human gut immunity. Nutr Rev 1998;56:S5–S8.

3 Mackie RI, Sghir A, Gaskins HR: Developmental microbial ecology of the neonatal gastrointestinal tract. Am J Clin Nutr 1999;69:1035S–1045S.

4 Gueimonde M, Laitinen K, Salminen S, Isolauri E: Breast-milk: a source of bifidobacteria for infant gut development and maturation? Neonatology 2007;92: 64–66.

5 Satokari R, Grönroos T, Laitinen K, Salminen S, Isolauri E: *Bifidobacterium* and *Lactobacillus* DNA in the human placenta. Lett Appl Microbiol 2009;48: 8–12.

6 Jiménez E, Marín ML, Martín R, Odriozola JM, Olivares M, Xaus J, Fernández L, Rodríguez JM: Is meconium from healthy newborns actually sterile? Res Microbiol 2008;159:187–193.

7 Williams AM, Probert CS, Stepankova R, Tlaskalova-Hogenova H, Phillips A, Bland PW: Effects of microflora on the neonatal development of gut mucosal T cells and myeloid cells in the mouse. Immunology 2006;119:470–478.

8 Rautava S, Collado MC, Salminen S, Isolauri E: Probiotics modulate host-microbe interaction in the placenta and fetal gut: a randomized, double-blind, placebo-controlled trial. Neonatology 2012;102:178–184.

9 Collado MC, Isolauri E, Laitinen K, Salminen S: Effect of mother's weight on infant's microbiota acquisition, composition, and activity during early infancy: a prospective follow-up study initiated in early pregnancy. Am J Clin Nutr 2010;92:1023–1030.

10 Collado MC, Laitinen K, Salminen S, Isolauri E: Maternal weight and weight gain during pregnancy modify the immunomodulatory potential of breast milk. Pediatr Res 2012;72:77–85.

11 Kalliomäki M, Kirjavainen P, Eerola E, Kero P, Salminen S, Isolauri E: Distinct patterns of neonatal gut microflora in infants in whom atopy was and was not developing. J Allergy Clin Immunol 2001;107:129–134.

12 Collado MC, Donat E, Ribes-Koninckx C, Calabuig M, Sanz Y: Specific duodenal and faecal bacterial groups associated with paediatric coeliac disease. J Clin Pathol 2009;62:264–269.

13 Kalliomäki M, Collado MC, Salminen S, Isolauri E: Early differences in fecal microbiota composition in children may predict overweight. Am J Clin Nutr 2008;87:534–538.

14 Neu J, Rushing J: Cesarean versus vaginal delivery: long-term infant outcomes and the hygiene hypothesis. Clin Perinatol 2011;38:321–331.

15 Bäckhed F, Ding H, Wang T, Hooper LV, Koh GY, Nagy A, Semenkovich CF, Gordon JI: The gut microbiota as an environmental factor that regulates fat storage. Proc Natl Acad Sci USA 2004;101:15718–15723.

16 Cani PD, Amar J, Iglesias MA, Poggi M, Knauf C, Bastelica D, Neyrinck AM, Fava F, Tuohy KM, Chabo C, Waget A, Delmée E, Cousin B, Sulpice T, Chamontin B, Ferrières J, Tanti JF, Gibson GR, Casteilla L, Delzenne NM, Alessi MC, Burcelin R: Metabolic endotoxemia initiates obesity and insulin resistance. Diabetes 2007;56:1761–1772.

17 Bäckhed F, Manchester JK, Semenkovich CF, Gordon JI: Mechanisms underlying the resistance to diet-induced obesity in germ-free mice. Proc Natl Acad Sci USA 2007;104:979–984.

18 Ley RE, Bäckhed F, Turnbaugh P, Lozupone CA, Knight RD, Gordon JI: Obesity alters gut microbial ecology. Proc Natl Acad Sci U S A 2005;102:11070–11075.

19 Turnbaugh PJ, Ley RE, Mahowald MA, Magrini V, Mardis ER, Gordon JI: An obesity-associated gut microbiome with increased capacity for energy harvest. Nature 2006;444:1027–1031.

20 Angelakis E, Armougom F, Million M, Raoult D: The relationship between gut microbiota and weight gain in humans. Future Microbiol 2012;7:91–109.

21 Collado MC, Isolauri E, Laitinen K, Salminen S: Distinct composition of gut microbiota during pregnancy in overweight and normal-weight women. Am J Clin Nutr 2008;88:894–899.

22 Santacruz A, Collado MC, García-Valdés L, Segura MT, Martín-Lagos JA, Anjos T, Martí-Romero M, Lopez RM, Florido J, Campoy C, Sanz Y: Gut microbiota composition is associated with body weight, weight gain and biochemical parameters in pregnant women. Br J Nutr 2010;104:83–92.

23 Lawlor DA, Lichtenstein P, Fraser A, Långström N: Does maternal weight gain in pregnancy have long-term effects on offspring adiposity? A sibling study in a prospective cohort of 146,894 men from 136,050 families. Am J Clin Nutr 2011;94:142–148.

24 Perez PF, Doré J, Leclerc M, Levenez F, Benyacoub J, Serrant P, Segura-Roggero I, Schiffrin EJ, Donnet-Hughes A: Bacterial imprinting of the neonatal immune system: lessons from maternal cells? Pediatrics 2007;119:e724–e732.

25 Goldani HA, Bettiol H, Barbieri MA, Silva AA, Agranonik M, Morais MB, Goldani MZ: Cesarean delivery is associated with an increased risk of obesity in adulthood in a Brazilian birth cohort study. Am J Clin Nutr 2011;93:1344–1347.

26 Barros FC, Matijasevich A, Hallal PC, Horta BL, Barros AJ, Menezes AB, Santos IS, Gigante DP, Victora CG: Cesarean section and risk of obesity in childhood, adolescence, and early adulthood: evidence from 3 Brazilian birth cohorts. Am J Clin Nutr 2012;95:465–470.

27 Hunt KM, Foster JA, Forney LJ, Schütte UM, Beck DL, Abdo Z, Fox LK, Williams JE, McGuire MK, McGuire MA: Characterization of the diversity and temporal stability of bacterial communities in human milk. PLoS One 2011;6:e21313.

28 Cabrera-Rubio R, Collado MC, Laitinen K, Salminen S, Isolauri E, Mira A: The human milk microbiome changes over lactation and is shaped by maternal weight and mode of delivery. Am J Clin Nutr 2012;96:544–551.

29 Hoppu U, Isolauri E, Laakso P, Matomäki J, Laitinen K: Probiotics and dietary counselling targeting maternal dietary fat intake modifies breast milk fatty acids and cytokines. Eur J Nutr 2012;51:211–219.

30 Luoto R, Kalliomäki M, Laitinen K, Delzenne NM, Cani PD, Salminen S, Isolauri E: Initial dietary and microbiological environments deviate in normal-weight compared to overweight children at 10 years of age. J Pediatr Gastroenterol Nutr 2011;52:90–95.

31 Luoto R, Kalliomäki M, Laitinen K, Isolauri E: The impact of perinatal probiotic intervention on the development of overweight and obesity: follow-up study from birth to 10 years. Int J Obes (Lond) 2010;34:1531–1537.

32 Laitinen K, Poussa T, Isolauri E, Nutrition, Allergy, Mucosal Immunology and Intestinal Microbiota Group: Probiotics and dietary counselling contribute to glucose regulation during and after pregnancy: a randomised controlled trial. Br J Nutr 2009;101:1679–1687.

33 Ilmonen J, Isolauri E, Poussa T, Laitinen K: Impact of dietary counselling and probiotic intervention on maternal anthropometric measurements during and after pregnancy: a randomized placebo-controlled trial. Clin Nutr 2011;30:156–164.

34 Luoto R, Laitinen K, Nermes M, Isolauri E: Impact of maternal probiotic-supplemented dietary counselling on pregnancy outcome and prenatal and postnatal growth: a double-blind, placebo-controlled study. Br J Nutr 2010;103:1792–1799.

Raakel Luoto, MD, PhD
Department of Paediatrics, Satakunta Central Hospital
Sairaalantie 3
28500 Pori (Finland)
E-Mail raakel.luoto@utu.fi

Guarino A, Quigley EMM, Walker WA (eds): Probiotic Bacteria and Their Effect on Human Health and Well-Being.
World Rev Nutr Diet. Basel, Karger, 2013, vol 107, pp 103–121 (DOI: 10.1159/000345750)

Metabolic Syndrome and Obesity in Adults

Susan E. Power[a, b] · Gerald F. Fitzgerald[a, b] · Paul W. O'Toole[a, b] ·
R. Paul Ross[b, c] · Catherine Stanton[b, c] · Eamonn M.M. Quigley[b, d] ·
Eileen F. Murphy[e]

[a]Department of Microbiology, [b]Alimentary Pharmabiotic Centre, and [d]Department of Medicine,
University College Cork, Cork, [c]Teagasc Food Research Centre, Moorepark, Fermoy, Co. Cork, and
[e]Alimentary Health Ltd., Cork, Ireland

Abstract

The relatively recent discovery that changes in the composition and metabolic activity of the gut microbiota are associated with obesity and related disorders has led to an explosion of interest in this now distinct research field. In the following chapter, we discuss the current evidence related to how the modulation of gut microbial populations might have beneficial effects with respect to controlling obesity. A number of studies in both animals and humans have shown that the composition of the gut microbiota is significantly altered in obesity and diabetes. Strategies including specific functional foods, probiotics, and prebiotics have the potential to favorably influence host metabolism by targeting the gut microbiota. Indeed, probiotics appear to be a promising approach to alter the host metabolic alterations linked to the changes in the gut microbiota. However, the mechanisms by which probiotics may impact on the development of obesity and metabolic health remain unclear and require further investigation.   Copyright © 2013 S. Karger AG, Basel

The prevalence of obesity and its associated metabolic disorders has increased substantially over recent decades to a point where they have reached epidemic levels worldwide. As such, obesity is a major public health issue and is associated with an increased risk of cardiovascular disease (CVD), type 2 diabetes mellitus (T2D), atherosclerosis, nonalcoholic fatty liver disease (NAFLD), and certain cancers. Obesity is a complex condition that results from an imbalance between energy intake and expenditure and appears to be influenced by a combination of genetic, lifestyle, and environmental factors. In this chapter the current evidence linking gut microbiota with the development of obesity and obesity-related metabolic diseases and the effects of interventions and probiotics, in particular, on the management and prevention of obesity will be discussed.

Gut Microbiota and Obesity

Animal Studies

The availability of animal models and the advent of new molecular, culture-independent techniques have together greatly enhanced our ability to study the gut microbiota and its relationship to obesity. Mice, in particular, offer a suitable model for conducting such experiments as their gut microbiota, by virtue of a predominance of Firmicutes (60–80%) and Bacteroidetes (20–40%), exhibits an overall pattern that is similar to that of humans [1]. Furthermore, germ-free (GF) mice are particularly useful for understanding the role of the gut microbiota in health and disease and the mechanisms underlying microbiota-associated alterations in host physiology. A landmark study by Bäckhed et al. [2] found that GF mice were leaner than their conventional (CONV) counterparts, even though they consumed more food. In addition, colonization of GF mice with the gut microbiota harvested from CONV mice led to an increase in total body fat despite a decrease in food intake. GF mice have also been shown to be protected against obesity while consuming a Western-style high-fat diet [3, 4], but the source and type of fat and carbohydrate may be an important factor [4]. In fact, Fleissner et al. [4] observed that GF mice fed a high-fat diet lacking sucrose were no longer protected against the development of diet-induced obesity, illustrating the complexity of the interactions between diet, microbes, and host adiposity.

There is mounting evidence that it is not simply the presence, but also the relative proportions of the major microbial divisions, within the microbiota that is associated with obesity. The development of obesity is associated with an increase in the ratio of Firmicutes to Bacteroidetes in both diet-induced obese [5, 6] and genetically obese *(ob/ob)* mice [1, 7]. Furthermore, a recent study showed that feeding a high-fat diet determined the composition of the gut microbiota independent of obesity in mice [8]. In that study, Hildebrandt et al. [8] showed, using the resistin-like molecule-β knockout mice (which are resistant to diet-induced obesity), that the high-fat diet itself, and not the obese state, mainly accounted for the observed changes in the gut microbiota composition. Other work showed that switching from a low-fat to a high-fat diet resulted in a rapid and dramatic shift in the structure of the gut microbiota in mice in a single day [5]. Murphy et al. [9] investigated the effects of a high-fat diet and genetic obesity (the *ob/ob* mouse model) on the gut microbiota over time. While no significant changes in the proportions of the different microbial groups were observed in the control lean mice over the 8-week period, a progressive increase in Firmicutes in both the high-fat-fed and *ob/ob* mice was reported, but reaching statistical significance in the former case only. Moreover, Bacteroidetes proportions decreased overtime in all groups, but again reached statistical significance only in the *ob/ob* mice [9]. These studies suggest that microbial adaptation to diet over time, and perhaps with age, is an important variable in the complex relationship between the composition of the microbiota and obesity, and should be considered in future studies.

Human Studies

In man the relationship between the gut microbiota and obesity is somewhat more unclear. The first study examining the qualitative changes of the gut microbiota in 12 human obese individuals was published by Ley et al. [10] in 2006. In agreement with results from animal studies, the authors observed an increased ratio of Firmicutes to Bacteroidetes in obese individuals compared to matched lean individuals. Interestingly, after weight loss (following a fat-restricted or carbohydrate-restricted low-calorie diet), the ratio of Bacteroidetes to Firmicutes approached a lean-type profile after 52 weeks [10]. Comparison of the fecal microbiota of monozygotic and dizygotic twins who were either lean or obese also revealed that obesity was associated with a reduced representation of Bacteroidetes, reduced bacterial diversity, and increased proportions of Actinobacteria. Notably, no significant differences in the proportions of Firmicutes were apparent between lean and obese subjects [11]. Although a similar reduction of Bacteroidetes in obese subjects has been confirmed in another study [12], the concept that Bacteroidetes are reduced in obese subjects has been largely contradicted by other studies [13–16].

Duncan et al. [13] detected no differences in Bacteroidetes proportions between obese and nonobese individuals and no significant changes in the proportions of Bacteroidetes were detected in obese subjects following weight loss. Furthermore, Schwiertz et al. [14] reported a decrease in the ratio of Firmicutes to Bacteroidetes in obese human adults compared with lean controls. Similarly, Zhang et al. [15] reported that overweight individuals harbored more Bacteroidetes than normal-weight individuals. They showed that a subgroup of Bacteroidetes (Prevotellaceae) was significantly enriched in obese individuals. Moreover, surgical treatment for these morbidly obese subjects (gastric bypass) altered the gut microbiota toward an increase in γ-Proteobacteria (members of Enterobacteriaceae) and a proportional decrease in Firmicutes [15]. More recently, Zupancic et al. [16] found no association between the Bacteroidetes-Firmicutes ratio and any metabolic syndrome trait in an Old Order Amish sect. However, 22 bacterial species were found to be either positively or inversely correlated with obesity and metabolic syndrome traits, indicating that specific members of the gut microbiota may play a role in these metabolic derangements [16]. The conflicting results observed in these human studies emphasize the complexity of the relationship between gut microbiota and obesity. Further work is required to identify if obesogenic components of the gut microbiota exist. Moreover, the potential influence of confounding factors, such as diet, highlights the need for well-controlled large clinical trials to identify specific proximate microbiota-related biomarkers of risk for obesity and metabolic dysregulation.

Mechanisms Linking the Gut Microbiota to Obesity

Host Factors. The gut microbiota may be involved in the regulation of fat storage and composition. *Fiaf* (fasting-induced adipocyte factor) is involved in regulating fat storage by inhibiting lipoprotein lipase, while also promoting fatty acid release by inducing peroxisome proliferator-activated receptor coactivator [2]. It has been suggested that GF mice are resistant to obesity through a mechanism involving *Fiaf* in the intestine [2]. Indeed, GF *Fiaf* knockout (*Fiaf–/–*) mice gained significantly more weight than their GF wild-type littermates when fed a Western diet [3]. However, a more recent study found that while intestinal *Fiaf* mRNA was increased in GF mice, there was no increase in plasma protein levels compared to CONV mice [4], suggesting that the *Fiaf* mechanism may not be universally associated with intestinal microbiota-related adiposity. Other work has suggested that the gut microbiota and its products affect host energy regulation acting through mechanisms involving adenosine monophosphate-activated protein kinase [3]. Furthermore, the microbiota may also influence food intake and energy expenditure of the host [2, 4, 17]. These studies suggest that a number of host-related factors may be involved in the regulation of adiposity by the gut microbiota.

Energy Harvesting. Colonization of GF mice with an 'obese microbiota' (i.e. microbiota from an obese animal) leads to a significantly greater increase in adiposity compared to GF mice colonized with a 'lean microbiota' [6, 7]. The gut microbiota may affect obesity by increasing energy harvesting from indigestible polysaccharides. It is estimated that around 20–60 g of undigested dietary carbohydrates reach the colon on a daily basis [18] where they are broken down by the gut microbiota to short-chain fatty acids (SCFAs), thereby contributing energy to the host (estimated to account for 5–10% of daily energy intake) [19]. Metagenomic analysis of *ob/ob* cecal microbiota revealed that this obese microbiome had an increased capacity to harvest energy from the diet as it was enriched with bacterial genes encoding enzymes for the utilization and fermentation of dietary fibers [7]. Moreover, the *ob/ob* cecum had increased concentrations of the major fermentation end-products butyrate and acetate, and *ob/ob* mice had significantly less energy remaining in their feces relative to their lean counterparts. It has also been shown that *ob/ob* mice harbor more methanogenic Archaea, which would be expected to increase the efficiency of bacterial fermentation by removing one of its end-products, H_2 [7]. Indeed, cocolonization of GF mice with *Methanobrevibacter smithii* (the main methanogenic archaeon in the human intestine) and *Bacteroides thetaiotaomicron* (a major saccharolytic member of the human intestine) increased polysaccharide fermentation efficiency and adiposity compared with mice colonized with either organism alone [20]. However, age may be an important factor in determining the potential of the gut microbiota to extract energy. Although *ob/ob* mice had less fecal energy and higher cecal concentrations of SCFAs compared to lean mice at 7 weeks of age, these patterns did not persist to the ages of 11 and 15 weeks [9]. In humans, SCFA levels were shown to be significantly increased and the proportion of individual SCFAs changed in favor of propionate in obese individuals compared to their lean counterparts [14]. A recent study reported that an altered nutrient load induced rapid changes in the composition of the human gut microbiota which were directly associated with stool energy loss in lean individuals. In this case, a 20% increase in Firmicutes and a corresponding decrease in Bacteroidetes were associated with an increased energy harvest of approximately 150 kcal [21].

Inflammation. Recent research has suggested that altered function of the innate immune system may promote the development of metabolic syndrome through a mechanism involving the gut microbiota. Obesity is generally associated with low-grade, chronic, systemic inflammation [22]. In fact, in murine models of obesity (genetic and diet-induced), the adipose tissue exhibits increased expression and content of proinflammatory cytokines [23, 24]. Cani et al. [25] demonstrated that excess dietary fat facilitates the absorption of the highly proinflammatory bacterial component lipopolysaccharide (LPS), a Toll-like receptor 4 (TLR-4) ligand, from the intestine. Mice fed a high-fat diet for 4 weeks had plasma LPS levels 2 or 3 times higher than normal, which was defined as 'metabolic endotoxemia'. Indeed, the direct infusion of LPS mimicked the physiological effects of the high-fat diet. Moreover, the effects of the high-fat diet were ameliorated in mice lacking CD14, a component of the TLR-4 receptor complex [25]. In a follow-up study, the gut microbiota was implicated in metabolic endotoxemia through the use of oral antibiotics which significantly decreased the levels of gut bacteria and systemic metabolic endotoxemia [26]. In addition, while plasma LPS concentrations correlated negatively with *Bifidobacterium* spp. [26], feeding mice a prebiotic (oligofructose) restored the number of bifidobacteria and reduced the impact of high-fat diet-induced metabolic endotoxemia [27]. This was associated with a reduction in gut permeability, improved tight junction integrity, and increased expression of glucagon-like peptide 2 [28].

Short-Chain Fatty Acids. In addition to their other physiological roles, recent data also suggest that SCFAs can have a role as signaling molecules [29] by acting as ligands for at least two G protein-coupled receptors (GPR41 and GPR43) on gut enteroendocrine cells [29, 30]. A recent study has shown that GPR43 knockout mice are protected from high-fat-diet-induced obesity and its consequences on glucose and lipoprotein metabolism [30]. Moreover, CONV GPR41 knockout (GPR41–/–) mice and GF GPR41–/– mice colonized with a model fermentative community (*B. thetaiotaomicron* and *M. smithii*) were significantly leaner than their wild-type counterparts, despite similar food intakes. However, there was no difference between GF wild-type and GF GPR41–/– mice indicating that the effects of GPR41 are dependent on the gut microbiota. Moreover, GPR41 deficiency was also associated with reduced expression of peptide YY, an enteroendocrine cell-derived hormone that normally inhibits gut motility, reduces intestinal transit rate, and increases harvest of energy (SCFAs) from the diet [29]. Therefore, these studies support the theory that intestinal metabolites (SCFAs) may also act as metabolic regulators in the host.

Flagellum and Toll-like receptor 5. TLR-5 is highly expressed in the intestinal mucosa and is involved in arbitrating immune responses through recognition of bacterial flagellum. Vijay-Kumar et al. [31] reported that TLR-5-knockout mice exhibited hyperphagia and demonstrated many of the features associated with the metabolic syndrome, including increased body mass and visceral fat deposition, dyslipidemia, hypertension, and insulin resistance. Moreover, these metabolic changes correlated with changes in the composition of the gut microbiota which were mainly at a species rather than phylum level. Indeed, transplantation of the microbiota from TLR-5-knockout mice to wild-type GF mice conferred many aspects of the TLR-5-knockout phenotype, including hyperphagia, obesity, hyperglycemia, and insulin resistance.

In summary, studies suggest that the gut microbiota and their metabolic products can influence obesity but the relationship between diet, the gut microbiota, and obesity is complex. Changes in diet not only lead to obesity by caloric intake per se, but also affect the composition of the gut microbiota which may in turn lead to obesity through several mechanisms (e.g. regulation of energy balance and low-grade inflammation). The question as to whether specific populations are responsible for obesity or just indicators (i.e. are these populations a 'cause' or an 'effect') remain unanswered. Taken together, these findings indicate that therapeutic manipulation of the microbiota may be a useful strategy in the prevention or management of obesity.

Gut Microbiota and Obesity-Related Metabolic Disease

The role of the gut microbiota in obesity has been studied in detail and the majority of the published literature describes the disparities between gut microbiota in obese compared to lean subjects. T2D, CVD, and NAFLD are generally considered as consequences of obesity. Though relationships between the microbiota and these disorders are less well studied than in obesity, per se, inferences regarding the role of the gut microbiota in these obesity-related conditions are starting to be made.

Gut Microbiota and Type 2 Diabetes

Low-grade inflammation is a common feature of T2D [22], and LPS is a possible initiator of metabolic impairment [25]. Indeed, subjects with T2D have higher LPS levels than subjects without diabetes [32, 33], and the infusion of LPS in mice induces low-grade chronic inflammation and features of early onset of metabolic diseases such as visceral fat deposition, glucose intolerance, and hepatic insulin resistance [25]. Moreover, the development of T2D in mice has been associated with increased endotoxemia along with increased gut permeability and systemic/adipose tissue inflammation [34].

However, it has been shown that not all genetically identical mice develop diabetes when fed a high-fat, carbohydrate-free diet, indicating that the metabolism of each phenotype is different despite identical genetic backgrounds and diets [34, 35]. Serino et al. [34] identified a gut microbial profile specific to diabetes-sensitive and diabetes-resistant metabolic phenotypes and found that an increased Bacteroidetes-Firmicutes ratio and a reduction in the Lachnospiraceae family and *Oscillibacter* genus were associated with the diabetic phenotype. Furthermore, modulation of the gut microbiota of these high-fat fed mice with dietary fibers prevented the occurrence of the diabetic phenotype and resulted in a specific microbial signature. Similarly, modulation of the gut microbiota with antibiotics has been shown to reverse insulin resistance in *ob/ob* mice independent of obesity [36]. It has also been demonstrated that GF mice are resistant to high-fat, diet-induced insulin resistance [17].

A small number of human, clinical studies have also reported on the relationship between gut microbiota composition and T2D. Indeed, there is increasing evidence which suggests that T2D can be associated with the gut microbiota, irrespective of the presence of obesity [37, 38]. The proportions of phylum Firmicutes and the class Clostridia were shown to be reduced while the class β-Proteobacteria was enriched in diabetic subjects compared to nondiabetic controls. Interestingly, the ratios of Bacteroidetes to Firmicutes, as well as the ratios of *Bacteroides-Prevotella* group to the *Clostridium coccoides-Eubacterium rectale* group, correlated positively and significantly with plasma glucose concentration, but not with BMI [37]. Other studies have found increased *Bacteroides* and reduced *Prevotella, Bifidobacterium, Bacteroides vulgatus* [38], and *Faecalibacterium prausnitzii* proportions associated with diabetes [39]. Recently, deep shotgun sequencing of the gut microbial DNA from 345 Chinese individuals identified approximately 60,000 T2D-associated markers, a decrease in the abundance of some universal butyrate-producing bacteria, and an increase in various opportunistic pathogens, as well as an enrichment of other microbial functions conferring sulfate reduction and oxidative stress resistance. An analysis of 23 additional individuals demonstrated that these gut microbial markers might be useful for classifying T2D [40]. Furthermore, a recent study has shown that infusion of the gut microbiota from lean donors to male recipients with metabolic syndrome for 6 weeks resulted in increased insulin sensitivity of the recipients along with increased levels of butyrate-producing gut microbiota [41]. These latter two studies highlight the role of SCFA, in particular butyrate in the relationship between the gut microbiota and T2D.

Gut Microbiota and Cardiovascular Disease
A link between the gut microbiota and CVD has recently been established following the discovery of a microbial-dependent pathway for the metabolism of dietary phospholipids which generates metabolites that are proatherosclerotic after absorption and hepatic metabolism [42]. The gut microbiota metabolizes dietary phosphatidylcholine (present in high-fat foods) to trimethylamine, which is then transported to the liver and oxidized to trimethylamine-N-oxide (TMAO). Wang et al. [42] found that increasing levels of plasma TMAO, choline, and betaine was found to have dose-dependent relationships with the presence of CVD in a cohort of 1,876 men and women. Furthermore, dietary supplementation of atherosclerosis-prone mice with choline, TMAO, or betaine promoted upregulation of multiple macrophage scavenger receptors linked to atherosclerosis, and dietary supplementation with choline or TMAO promoted atherosclerosis. This study suggests opportunities for the development of novel therapeutic strategies for the treatment for atherosclerosis.

Furthermore, it has been shown that GF mice are resistant to high-fat, diet-induced dyslipidemia and elevated proinflammatory markers compared to their CONV counterparts. Moreover, there are unexpected changes in cholesterol metabolism, including reduced hypercholesterolemia, a moderate increase of hepatic cholesterol, an increase in cholesterol excretion, and an upregulation of cholesterol biosynthesis gene [17].

Low-grade inflammation is a common feature of CVD [22] and metabolic endotoxemia has also been linked to its development. In a 5-year prospective study (the Bruneck study), it was demonstrated that subclinical endotoxemia constitutes a strong risk factor for the development of carotid atherosclerosis, particularly among smokers [43].

The relationship between both the oral and gut microbiota and atherosclerosis has recently been investigated. Although no relationship was found between plaque microbiota and gut microbiota in affected patients, the abundance of certain bacteria (*Veillonella* and *Streptococcus*) in atherosclerotic plaque correlated with their abundance in the oral cavity [44].

Gut Microbiota and Nonalcoholic Fatty Liver Disease
Although there is a relative paucity of data, the microbiota may also be linked to the development and progression of NAFLD and nonalcoholic steatohepatitis (NASH). It has been shown that GF mice are resistant to high-fat-diet-induced hepatic steatosis [17] and colonization of GF mice with the gut microbiota harvested from CONV mice increased liver fat and liver triglyceride content [2]. Moreover, NAFLD and NASH have been associated with increased intestinal permeability, as well as small intestinal bacterial overgrowth [45]. Several molecular mechanisms relevant to the involvement of the microbiota in NAFLD/NASH have been proposed. The increased production of SCFA by the microbiota would provide more energy to the liver [3, 7]. The microbiota has also been shown to stimulate hepatic triglyceride production through suppression of *Fiaf* [2], and its role in choline metabolism has also been implicated as one of the mechanisms which contribute to NAFLD [42, 46]. As liver disease is associated with a state of inflammation, LPS is also likely to be involved in NAFLD/NASH. Indeed, increased levels of LPS have been observed in NAFLD patients [47]. More recently, it has been shown that genetic inflammasome deficiency and its related dysbiosis result in abnormal accumulation of bacterial products in the portal circulation and drive progression of NAFLD/NASH [48].

In summary, the gut microbiota represents an important source of metabolic variability in the host and may also play a role in the development of metabolic disorders such as T2D, NAFLD, and CVD, perhaps even independently of obesity. Increased knowledge of the mechanisms involved in the interactions between the microbiota and its host will aid in the development of treatments for metabolic disease.

Probiotics: Role in Obesity and Metabolic Syndrome

The demonstration of the possible role of gut microbiota in the pathogenesis of obesity and related metabolic diseases has led to the development of different strategies to modulate the gut microbiota and hence ameliorate host metabolism. In this respect, there are differing approaches to modulate the gut microbiota. Specific bacterial species or classes may be reduced by antibiotics. Studies have shown that treatment of both *ob/ob* and diet-induced obese mice with broad spectrum antibiotics leads to changes in the composition of the gut microbiota as well as an improvement in metabolic abnormalities, including improved insulin resistance, fasting glycemia, glucose tolerance, and a reduction in both metabolic endotoxemia and body weight gain [26, 36, 49].

In contrast to reducing specific bacterial species through the use of antibiotics, certain beneficial bacteria may be increased either by pre- or probiotics. As discussed previously, Cani et al. [27] found that the number of cecal *Bifidobacterium* spp. was inversely correlated with the development of fat mass and glucose intolerance, as well as LPS levels, in high-fat diet fed mice. Prebiotics containing oligofructose were found to completely restore *Bifidobacterium* species, normalize plasma endotoxin levels, and lead to improved glucose tolerance and glucose-induced insulin secretion when administered to high-fat fed mice. Similarly, Serino et al. [34] recently showed that the prebiotic glucose oligosaccharide reversed most aspects of the high-fat induced diabetic phenotype in mice.

The application of probiotics as a potential therapy in the management of obesity and its associated metabolic disorders has also received particular attention and has been explored in a number of studies.

Animal Studies

A number of experimental studies suggest that the probiotic approach can be used to alter biomarkers of metabolic disease (table 1). Indeed, specific *Lactobacillus* and *Bifidobacterium* strains have been shown to reduce body weight and fat mass [50–57] and alter lipid profiles [57–61] in murine models, although the effect is likely to be strain dependent [62]. For example, milk products fermented with *L. gasseri* SBT2055 were shown to have a possible role in the regulation of adipose tissue growth by reducing adipocyte size, serum leptin [63], and fatty acid absorption [59], possibly, through an anti-inflammatory mechanism [64]. In contrast, other work has shown an increase in body weight with the administration of *L. fermentum* (CIP 102980) to chickens [65]. Indeed, a recent debate regarding the role of probiotics in metabolic disease prompted a proposal that these 'friendly' bacteria may play a role in the development of obesity. Raoult [66] suggested that since probiotics are used as growth promoters in the farm industry, there is a possibility that probiotics may also promote weight gain in humans. However, this view has largely been rejected by other researchers in the field [67].

Probiotic strains may also be able to modify the fat composition of host tissues. Wall et al. [68] showed that the administration of a conjugated linoleic acid-producing *B. breve* strain, in combination with linoleic acid as a substrate, led to a modulation of the fatty acid composition of mice and pigs, including significantly elevated concentrations of hepatic and adipose tissue content of *cis-9*, *trans-11* conjugated linoleic acid, and omega-3 fatty acids, along with reduced levels of the proinflammatory cytokines TNF-α and IFN-γ.

Table 1. Effects of probiotic consumption on biomarkers of metabolic disease in animal models

Author	Probiotic	Model	Duration	Outcome
Aronsson et al. [51]	*L. paracasei* ssp. *paracasei* F-19 (2×10^9 CFU/g feed)	male C57BL/6J mice – high-fat diet	10 weeks	↓ body fat ↑ circulating *Fiaf*
Hamad et al. [59]	skim milk fermented with *L. gasseri* SBT2055 (6×10^7 CFU/g diet)	lean and obese Zucker rats	4 weeks	↓ serum total and HDL cholesterol ↑ excretion of fecal fatty acids and total neutral fecal sterols *lean rats:* ↓ mesenteric adipose tissue weight ↓ adipocyte sizes ↓ serum leptin *obese rats:* ↑ number of smaller adipocytes in the subcutaneous adipose tissue
	skim milk fermented with *L. gasseri* SBT2055 (6×10^7 CFU/g)	male Sprague-Dawley rats (with permanent cannulation of thoracic duct)	1 week	↓ maximum transport rate of TG and phospholipids
Ji et al. [56]	*L. rhamnosus* GG (1×10^8 CFU/day) or *L. sakei* NR28 (1×10^8 CFU/day)	male C57BL/6J mice	3 weeks	↓ epididymal fat ↓ obesity related biomarkers such as FAS, ACC, and SCD-1 from the liver
Kadooka et al. [64]	yoghurt containing *L. gasseri* SBT2055 (5×10^8 CFU/g)	male Sprague-Dawley rats	4 weeks	inhibition of enlargement of visceral adipocytes prevention of upregulation of sICAM-1 (inflammatory marker that is elevated in obesity)
Kang et al. [50]	*L. gasseri* BNR17 (10^9 CFU/0.5 ml – administered twice daily)	male Sprague-Dawley rats – high carbohydrate diet	12 weeks	↓ weight gain ↓ fat pad mass ↔ serum total, HDL, and LDL cholesterol ↔ TG
Kondo et al. [57]	*B. breve* B-3 (10^8 or 10^9 CFU/day)	male C57BL/6J mice – high-fat DIO	8 weeks	↓ body weight and epididymal fat ↓ serum levels of total cholesterol, fasting glucose, and insulin ↑ expression of genes related to fat metabolism and insulin sensitivity in gut and epididymal fat tissue
Lee et al. [54]	*L. rhamnosus* PL60 (1×10^7 CFU/day or 1×10^9 CFU/day)	male C57BL/6J mice – high-fat DIO	8 weeks	↓ weight gain ↓ white adipose tissue (epididymal and perirenal) ↑ apoptotic signals and UCP-2 levels in adipose tissue
Lee et al. [53]	*L. plantarum* PL62 (1×10^7 CFU/day or 1×10^9 CFU/day)	male C57BL/6J mice – high-fat DIO	8 weeks	↓ weight of epididymal, inguinal, mesenteric, and perirenal white adipose tissue ↓ blood levels of glucose ↓ body weight
Ma et al. [71]	#VSL probiotics (mixture of bifidobacteria, lactobacilli, and *Streptococcus thermophilus*) (1.5×10^9 CFU/day)	male C57BL/6J mice – high-fat, diet-induced hepatic steatosis and insulin resistance	4 weeks	↑ hepatic NKT cell depletion ↓ weight improved insulin resistance improved steatosis
Martin et al. [61]	*L. paracasei* NCC2461 or *L. rhamnosus* NCC4007 (1×10^8 CFU/day)	female germ-free C3H mice – colonized with human baby flora	2 weeks	↓ cecal acetate and butyrate ↑ cecal isobutyrate and isovalerate with *L. paracasei* ↓ plasma LDL- and VLDL cholesterol ↑ TG

Table 1. Continued

Author	Probiotic	Model	Duration	Outcome
Naito et al. [72]	L. casei strain Shirota YIT 9029 (heat killed)	male C57BL/6J mice – high-fat DIO	4 weeks	improves insulin resistance and glucose intolerance ↓endotoxemia
Sato et al. [63]	skim milk fermented with L. gasseri SBT2055 (6×10^7 CFU/g diet)	male Sprague-Dawley rats	4 weeks	↓ adipocyte size in mesenteric white adipose tissue ↑number of small adipocytes from mesenteric and retroperitoneal adipose tissues ↓ serum leptin
Tabuchi et al. [70]	L. rhamnosus GG	male Wistar rats (streptozotocin-induced diabetes)	9 weeks	↓glycosylated hemoglobin improved glucose tolerance
Takemura et al. [55]	L. plantarum LP14 (1×10^8 CFU/mouse – administered intragastrically)	female C57BL/6 mice – high-fat diet	11 weeks	↓ adipocyte size ↓ white adipose tissue weight ↓ serum total cholesterol ↓ serum leptin
Tanida et al. [52]	L. paracasei ST11 (1×10^9 CFU/2 ml water)	male Wistar rats – high-fat DIO	11 weeks	↓ weight gain ↓ abdominal fat tissue weight
Wall et al. [68]	B. breve NCIMB 702258 (10^9 microorganisms/day) + linoleic acid	healthy BALB/c mice, SCID mice and weaning pigs	8 weeks	↑ hepatic and adipose tissue content of cis-9, trans-11 CLA, DHA, and EPA ↓ splenocytes production of TNF-α, IL-6, and IFN-γ
Xiao et al. [58]	milk product fermented with B. longum BL-1 (approx. 4×10^8 CFU/g)	male Sprague-Dawley rats	3 weeks	↓ serum total and LDL cholesterol ↓ TG
Yadav et al. [60]	dahi product containing L. acidophilus and L. casei	male Wistar rats (high-fructose-induced diabetes)	8 weeks	↓ plasma glucose, glycosylated hemoglobin and insulin and liver glycogen ↓ total, LDL and VLDL cholesterol ↓ TG and free fatty acids ↓ thiobarbituric acid-reactive substances ↑ reduced glutathione in liver and pancreas
Yadav et al. [69]	dahi product containing Lb. acidophilus and L. casei (7.3×10^9 CFU/g --15 g/day/rat)	male Wistar rats (streptozotocin-induced diabetes)	4 weeks	↓ oxidative damage in pancreatic tissues by inhibiting the lipid peroxidation and formation of nitric oxide; preserved antioxidant pool such as glutathione content and activities of super oxide dismutase, catalase and glutathione peroxidase
Yin et al. [62]	Bifidobacteria strains L66-5, L75-4, M13-4 and FS31-12	male Sprague-Dawley rats; high-fat diet	6 weeks	the response to energy metabolism was strain dependent: M13-4 ↑ body weight gain L66-5 ↓ body weight gain all 4 strains ↓ serum and liver TG

ACC = Acetyl-CoA carboxylase; CFU = colony forming units; CLA = conjugated linoleic acid; DHA = docosahexaenoic acid; DIO = diet-induced obesity; EPA = eicosapentaenoic acid; FAS = fatty acid synthase; NKT = natural killer T; SCD-1 = stearoyl-CoA desaturase-1; SCID = severe combined immunodeficient; sICAM-1 = soluble intercellular adhesion molecule-1; TG = triglycerides; UCP-2 = uncoupling protein-2.

In rat models of diabetes (high fructose-/streptozotocin-induced diabetes), specific *Lactobacillus* strains can improve such markers of diabetes as hyperinsulinemia, hyperglycemia, glucose intolerance, dyslipidemia, and oxidative stress [60, 69, 70]. Moreover, the probiotic product #VSL has been shown to improve insulin resistance, hepatic natural killer T cell depletion, and hepatic steatosis in diet-induced obese mice [71], while *L. casei* Shirota has been demonstrated to improve insulin resistance and glucose intolerance as well as reduce endotoxemia [72].

Human Studies
Only a limited number of human clinical trials have assessed the effects of probiotic intake on metabolic and obesity biomarkers (table 2).

Kadooka et al. [73] carried out a double-blind, randomized, placebo-controlled intervention trial among subjects (n = 87) with an increased BMI and abdominal visceral fat area. Subjects received either a fermented milk product containing *L. gasseri* LG2055 or the same product without LG2055 over a period of 12 weeks. Abdominal fat area was determined by computed tomography. In subjects receiving the probiotic, abdominal visceral and subcutaneous fat areas decreased significantly from baseline (by an average of 4.6 and 3.3%, respectively). Body weight, BMI, and hip and waist measures also decreased significantly, while none of those parameters changed significantly in the control group.

A large study by Laitinen et al. [74] showed that the combination of dietary counseling and a probiotic intervention led to consistently improved glucose metabolism and insulin sensitivity in a cohort (n = 256) of healthy normoglycemic pregnant women. Moreover, the probiotics were shown to provide a greater glucose-lowering effect than dietary counseling alone. In a double-blind randomized controlled trial, Andreasen et al. [75] allocated male subjects (n = 45) with T2D, impaired or normal glucose tolerance to a 4-week treatment with either *L. acidophilus* NFCM or placebo. Insulin sensitivity was preserved in the *L. acidophilus*-fed group, whereas it decreased in the placebo group. However, there was no change in inflammatory markers over the 4-week period, indicating that this effect was not linked to the immune system.

A number of studies have also investigated the cholesterol lowering effects of probiotic strains. Bukowska et al. [76] investigated the role of 'Pro viva' (a food product containing a fermentable oat fraction and the probiotic *L. plantarum* 299v) supplementation on the atherogenic markers LDL cholesterol and fibrinogen in a double-blind cross-over study among 30 male subjects. Six weeks of treatment resulted in a reduction in levels of both markers. In a subsequent study, Naruszewicz et al. [77] investigated the influence of *L. plantarum* in a double-blind randomized controlled trial involving 36 smokers and found that the 6-week treatment reduced systolic blood pressure along with several proatherogenic markers, including plasma concentrations of fibrinogen, IL-6, and F2-isoprostanes (markers of lipid oxidant stress). Moreover, *L. plantarum* administration markedly decreased the adherence of monocytes to native and TNF-activated endothelial cells. Similarly, Kullissar et al. [78] investigated the

Table 2. Effects of probiotic consumption on biomarkers of metabolic disease in human subjects

Author	Probiotic	Experimental design	Subjects (mean age)	Duration	Outcome
Andreasen et al. [75]	*L. acidophilus* NCFM (1×10^{10} CFU/day)	double-blind RCT – 2 arms (1) probiotic (2) placebo	45 males T2D, impaired or normal glucose tolerance (mean age 55.6 years)	4 weeks	insulin sensitivity was preserved in the probiotic group but ↓ in the placebo group; ↔ inflammatory markers or systemic immune response in either group
Bukowska et al. [76]	Pro viva – a food product containing fermentable oatmeal with *L. plantarum* 299v (~5×10^7 CFU/ml – 200 ml administered daily)	double-blind RCT – 2 arms (1) Pro viva product (2) rose-hip drink (control)	30 healthy males with moderately elevated serum cholesterol (mean age 42.6 years)	6 weeks	↓ total and LDL cholesterol ↓ fibrinogen ↔ triglycerides, HDL cholesterol, glucose levels, and BMI
Greany et al. [82]	*L. acidophilus* DDS-1, *B. longum* UABL-14 (10^9 CFU) and oligofructose (10–15 g) per capsule – administered 3 times a day	Single-blind RCT- 2 arms (1) probiotic (2) placebo	55 normocholesterolemic males and females (age range: 18–36 years)	2 months (males) or 2 menstrual cycles (females)	↔ plasma concentrations of total cholesterol, HDL cholesterol, LDL cholesterol, and triglycerides
Kadooka et al. [73]	Fermented milk containing *L. gasseri* SBT2055	double-blind RCT – 2 arms (1) fermented milk + probiotic (2) fermented milk	87 subjects with an increased BMI (24.2–30.7) and abdominal visceral fat area (81.2-178.5 cm^2).	12 weeks	↓abdominal visceral and subcutaneous fat areas ↓ body weight, BMI, hip, and waist measurements
Kiessling et al. [79]	yoghurt containing *L. acidophilus* 145 (10^6–10^8 CFU/g), *B. longum* 913 (>10^5 CFU/g) and 1% oligofructose (300 g administered daily)	cross-over study (7 week period each) (1) control (for all women) (2) and (3) control or probiotic	29 healthy females –including normocholesterolemic and hypercholesterolemic subjects (mean age 34 years)	7 weeks	↑ HDL cholesterol ↓ LDL/HDL cholesterol ratio. ↔ total cholesterol and LDL cholesterol
Kullisaar et al. [78]	goat milk fermented with *L. fermentum* ME-3 (3×10^{11} CFU/day)	CT – 2 arms (1) goat milk + probiotic (n = 16) (2) goat milk (control; n = 5)	21 healthy males and females (mean age 50 years)	3 weeks	enhanced total antioxidative activity prolonged resistance of the lipoprotein fraction to oxidation ↓ levels of peroxidized lipoproteins, oxidized LDL, and 8-isoprostanes ↓glutathione redox ratio
Laitinen et al. [74]	*L. rhamnosus*GGand *B. lactis*Bb12 (1×10^{10} CFU/day each)	double-blind RCT - 3 arms (1) diet/probiotics (2) diet/placebo (3) control/placebo	256 healthy pregnant females (mean age 30 years)	from 1st trimester to the end of exclusive breastfeeding	fasting plasma glucose, serum insulin and HOMA were lowest in the diet/probiotic group during 3rd trimester of pregnancy and 12 months postpartum
Lewis and Burmeister [80]	*L. acidophilus* LA-1 (6×10^{10} CFU – 3 times a day)	cross-over study (6 week period each) – (1) probiotic (2) control	80 healthy males and females with elevated cholesterol (mean age 47 years)	6 weeks	↔ plasma lipids

Table 2. Continued

Author	Probiotic	Experimental design	Subjects (mean age)	Duration	Outcome
Naruszewicz et al. [77]	rose-hip drink containing *L. plantarum* 299v (5 × 10^7 CFU/ml – 400 ml administered daily)	double-blind RCT – 2 arms (1) rose-hip drink + probiotic (2) rose-hip drink	36 healthy male and female smokers (mean age 42.3 years)	6 weeks	↓systolic BP and fibrinogen ↓ F2-isoprostanes and IL-6 isolated monocytes showed significantly reduced adhesion to native and stimulated endothelial cells
Simons et al. [81]	PCC® *L. fermentum* (2 × 10^9 CFU/capsule – 2 capsules administered twice daily)	double-blind RCT – 2 arms (1) PCC® Lb. fermentum (2) placebo	46 healthy males and females with elevated serum cholesterol (mean age 51.5 years)	10 weeks	LDL cholesterol showed a modest downward trend for both probiotic and placebo no significant difference in other measurements between treatment arms (total cholesterol, HDL cholesterol, triglycerides

BP = Blood pressure; CFU = colony-forming units; CT = controlled trial; HOMA = homeostatic model assessment (method used to quantify insulin resistance and β-cell function); RCT = randomized controlled trial; ↓ = decreased; ↑ = increased; ↔ = unchanged.

effect of a probiotic product (goat milk fermented with *L. fermentum* ME-3) on oxidative stress markers (including proatherogenic markers) in 21 healthy subjects. Consumption of the fermented goats' milk altered both the prevalence and proportion of lactic acid bacteria species in the gut microbiota, and exhibited antioxidative and antiatherogenic effects in the healthy subjects. Furthermore, Kiessling et al. [79] demonstrated in a cross-over type study including 29 women that daily consumption of a yogurt supplemented with *L. acidophilus* 145, *B. longum* 913, and oligofructose increased serum concentration of HDL cholesterol and led to the desired improvement of the LDL/HDL cholesterol ratio compared with a control yoghurt. Xiao et al. [58] investigated the effect of consuming low-fat yoghurt containing *B. longum* BL-2 on the lipid profiles of 32 healthy males and showed a decline in serum total cholesterol after 4 weeks, particularly among subjects with moderate hypercholesterolemia.

Although many of these reports demonstrate a cholesterol-lowering effect, a number of studies have failed to show such an effect in humans. For example, Lewis and Burmeister [80] conducted a cross-over type study in 80 healthy volunteers with elevated cholesterol and found that supplementation with *L. acidophilus* had no significant effect on plasma total, LDL, and HDL cholesterol and triglycerides. Simons et al. [81] also demonstrated that the consumption of *L. fermentum* did not contribute to any significant lipid profile changes after its administration to healthy subjects with elevated cholesterol levels for 10 weeks. Similarly, the administration of a synbiotic product (*L. acidophilus*, *B. longum*, and oligofructose) for approximately 2 months had no significant effect on plasma cholesterol or triglycerides in normocholesteremic volunteers [82]. These contradictory findings are probably due to differences in experimental design and study population, as well as variations in probiotic strain and dosage.

Many questions still remain regarding the role of probiotic bacteria in obesity and metabolic disease, and current evidence is derived mainly from animal studies. Although these results from animal studies are interesting, the underlying mechanisms are unclear. Only a minority of human studies have examined the benefits of probiotic bacteria on biomarkers of metabolic disorders, and these have shown conflicting results. In addition, the majority of these trials have been relatively small-sized (<50 subjects) and of short duration (<10 weeks). The differences between studies may arise due to the use different bacterial strains. Indeed, the results from one probiotic strain cannot be extrapolated to another. Moreover, quite often different delivery systems (e.g. fermented dairy product vs. freeze-dried bacteria), different population groups (e.g. normocholesterolemic vs. hypercholesterolemic), and different administration dosages have been employed, rendering comparisons difficult.

Conclusions

Obesity and associated metabolic diseases are increasing in prevalence worldwide. Both experimental and clinical data have demonstrated that obesity is associated with changes in the gut microbiota. Nonetheless, many unanswered questions still remain. Researchers have yet to conclusively identify 'obesogenic' bacterial groups. Moreover, the question still remains whether the differences in the microbiota between lean and obese groups are the cause or the consequence of obesity or perhaps a combination of both. Future research is therefore challenged with the task of identifying beneficial bacteria which are capable of controlling adiposity and related metabolic disorders, as well as with investigating possible dietary approaches to promote these bacteria in order to improve metabolic health. Indeed, data is accumulating which suggest that manipulation of the microbiome using prebiotics or probiotics may reduce insulin resistance and fat accumulation. Although a number of animal studies have revealed interesting results, it is difficult to apply these findings to man. Only a limited number of small-sized clinical trials have demonstrated the promise for specific probiotic strains in beneficially modulating biomarkers of metabolic disease in humans. However, the potential implications of this advancing field are promising.

Acknowledgements

The Alimentary Pharmabiotic Centre is a research center funded by Science Foundation Ireland (SFI) through the Irish Government's National Development Plan. S.E. Power is supported by the Irish Research Council postgraduate scholarship Enterprise Partnership Scheme (in collaboration with Alimentary Health Ltd.) and by the Alimentary Pharmabiotic Centre (SFI grant No. 07/CE/B1368). This work was also supported by the Government of Ireland National Development Plan by way of a Department of Agriculture Food and Marine, and Health Research Board FHRI award to the ELDERMET project.

References

1 Ley RE, Bäckhed F, Turnbaugh P, Lozupone CA, Knight RD, Gordon JI: Obesity alters gut microbial ecology. Proc Natl Acad Sci USA 2005;102:11070–11075.

2 Bäckhed F, Ding H, Wang T, Hooper LV, Koh GY, Nagy A, Semenkovich CF, Gordon JI: The gut microbiota as an environmental factor that regulates fat storage. Proc Natl Acad Sci USA 2004;101:15718–15723.

3 Bäckhed F, Manchester JK, Semenkovich CF, Gordon JI: Mechanisms underlying the resistance to diet-induced obesity in germ-free mice. Proc Natl Acad Sci USA 2007;104:979–984.

4 Fleissner CK, Huebel N, Abd El-Bary MM, Loh G, Klaus S, Blaut M: Absence of intestinal microbiota does not protect mice from diet-induced obesity. Br J Nutr 2010;104:919–929.

5 Turnbaugh PJ, Ridaura VK, Faith JJ, Rey FE, Knight R, Gordon JI: The effect of diet on the human gut microbiome: a metagenomic analysis in humanized gnotobiotic mice. Sci Translat Med 2009;1:6ra14.

6 Turnbaugh PJ, Bäckhed F, Fulton L, Gordon JI: Diet-induced obesity is linked to marked but reversible alterations in the mouse distal gut microbiome. Cell Host Microbe 2008;3:213–223.

7 Turnbaugh PJ, Ley RE, Mahowald MA, Magrini V, Mardis ER, Gordon JI: An obesity-associated gut microbiome with increased capacity for energy harvest. Nature 2006;444:1027–1031.

8 Hildebrandt MA, Hoffmann C, Sherrill-Mix SA, Keilbaugh SA, Hamady M, Chen YY, Knight R, Ahima RS, Bushman F, Wu GD: High-fat diet determines the composition of the murine gut microbiome independently of obesity. Gastroenterology 2009;137:1716–1724.

9 Murphy EF, Cotter PD, Healy S, Marques TM, O'Sullivan O, Fouhy F, Clarke SF, O'Toole PW, Quigley EM, Stanton C, Ross PR, Doherty RM, Shanahan F: Composition and energy harvesting capacity of the gut microbiota: relationship to diet, obesity and time in mouse models. Gut 2010;59:1635–1642.

10 Ley RE, Turnbaugh PJ, Klein S, Gordon JI: Microbial ecology: human gut microbes associated with obesity. Nature 2006;444:1022–1023.

11 Turnbaugh PJ, Hamady M, Yatsunenko T, Cantarel BL, Duncan A, Ley RE, Sogin ML, Jones WJ, Roe BA, Affourtit JP, Egholm M, Henrissat B, Heath AC, Knight R, Gordon JI: A core gut microbiome in obese and lean twins. Nature 2009;457:480–484.

12 Armougom F, Henry M, Vialettes B, Raccah D, Raoult D: Monitoring bacterial community of human gut microbiota reveals an increase in Lactobacillus in obese patients and Methanogens in anorexic patients. PLoS One 2009;4:e7125.

13 Duncan SH, Lobley GE, Holtrop G, Ince J, Johnstone AM, Louis P, Flint HJ: Human colonic microbiota associated with diet, obesity and weight loss. Int J Obe 2008;32:1720–1724.

14 Schwiertz A, Taras D, Schafer K, Beijer S, Bos NA, Donus C, Hardt PD: Microbiota and SCFA in lean and overweight healthy subjects. Obesity 2010;18:190–195.

15 Zhang H, DiBaise JK, Zuccolo A, Kudrna D, Braidotti M, Yu Y, Parameswaran P, Crowell MD, Wing R, Rittmann BE, Krajmalnik-Brown R: Human gut microbiota in obesity and after gastric bypass. Proc Natl Acad Sci USA 2009;106:2365.

16 Zupancic ML, Cantarel BL, Liu Z, Drabek EF, Ryan KA, Cirimotich S, Jones C, Knight R, Walters WA, Knights D, Mongodin EF, Horenstein RB, Mitchell BD, Steinle N, Snitker S, Shuldiner AR, Fraser CM: Analysis of the gut microbiota in the Old Order Amish and its relation to the metabolic syndrome. PLoS One 2012;7:e43052.

17 Rabot S, Membrez M, Bruneau A, Gérard P, Harach T, Moser M, Raymond F, Mansourin R, Chou CJ: Germ-free C57BL/6J mice are resistant to high-fat-diet-induced insulin resistance and have altered cholesterol metabolism. FASEB J 2010;24:4948–4959.

18 Flint H, Scott K, Duncan S, Louis P, Forano E: Microbial degradation of complex carbohydrates in the gut. Gut Microbes 2012;3:1–18.

19 McNeil NI: The contribution of the large intestine to energy supplies in man. Am J Clin Nutr 1984;39:338–342.

20 Samuel BS, Gordon JI: A humanized gnotobiotic mouse model of host-archaeal-bacterial mutualism. Proc Natl Acad Sci USA 2006;103:10011–10016.

21 Jumpertz R, Le DS, Turnbaugh PJ, Trinidad C, Bogardus C, Gordon JI, Krakoff J: Energy-balance studies reveal associations between gut microbes, caloric load, and nutrient absorption in humans. Am J Clin Nutr 2011;94:58–65.

22 Wellen KE, Hotamisligil GS: Inflammation, stress, and diabetes. J Clin Invest 2005;115:1111–1119.

23 Hotamisligil GS, Shargill NS, Spiegelman BM: Adipose expression of tumor necrosis factor-alpha: direct role in obesity-linked insulin resistance. Science 1993;259:87–91.

24 Weisberg SP, McCann D, Desai M, Rosenbaum M, Leibel RL, Ferrante AW Jr: Obesity is associated with macrophage accumulation in adipose tissue. J Clin Invest 2003;112:1796–1808.

25 Cani PD, Amar J, Iglesias MA, Poggi M, Knauf C, Bastelica D, Neyrinck AM, Fava F, Tuohy KM, Chabo C, Waget A, Delmée E, Cousin B, Sulpice T, Chamontin B, Ferrières J, Tanti JF, Gibson GR, Casteilla L, Delzenne NM, Alessi MC, Burcelin R: Metabolic endotoxemia initiates obesity and insulin resistance. Diabetes 2007;56:1761–1772.

26 Cani PD, Bibiloni R, Knauf C, Waget A, Neyrinck AM, Delzenne NM, Burcelin R: Changes in gut microbiota control metabolic endotoxemia-induced inflammation in high-fat diet-induced obesity and diabetes in mice. Diabetes 2008;57:1470–1481.

27 Cani P, Neyrinck A, Fava F, Knauf C, Burcelin RG, Tuohy KM, Gibson GR, Delzenne NM: Selective increases of bifidobacteria in gut microflora improve high-fat-diet-induced diabetes in mice through a mechanism associated with endotoxaemia. Diabetologia 2007;50:2374–2383.

28 Cani PD, Possemiers S, Van de Wiele T, Guiot Y, Everard A, Rottier O, Geurts L, Naslain D, Neyrinck A, Lambert DM, Muccioli GG, Delzenne NM: Changes in gut microbiota control inflammation in obese mice through a mechanism involving GLP-2-driven improvement of gut permeability. Gut 2009;58:1091–1103.

29 Samuel BS, Shaito A, Motoike T, Rey FE, Backhed F, Manchester JK, Hammer RE, Williams SC, Crowley J, Yanagisawa M, Gordon JI: Effects of the gut microbiota on host adiposity are modulated by the short-chain fatty-acid binding G protein-coupled receptor, Gpr41. Proc Natl Acad Sci USA 2008;105:16767–16772.

30 Bjursell M, Admyre T, Göransson M, Marley AE, Smith DM, Oscarsson J, Bohlooly-Y M: Improved glucose control and reduced body fat mass in free fatty acid receptor 2-deficient mice fed a high-fat diet. Am J Physiol Endocrinol Metab 2011;300:E211–E220.

31 Vijay-Kumar M, Aitken JD, Carvalho FA, Cullender TC, Mwangi S, Srinivasan S, Knight R, Ley RE, Gewirtz: Metabolic syndrome and altered gut microbiota in mice lacking Toll-like receptor 5. Science 2010;328:228–231.

32 Pussinen PJ, Havulinna AS, Lehto M, Sundvall J, Salomaa V: Endotoxemia is associated with an increased risk of incident diabetes. Diabetes Care 2011; 34:392–397.

33 Creely SJ, McTernan PG, Kusminski CM, Da Silva N, Khanolkar M, Evans M, Harte, AL, Kumar S: Lipopolysaccharide activates an innate immune system response in human adipose tissue in obesity and type 2 diabetes. Am J Physiol Endocrinol Metab 2007;292:E740–E747.

34 Serino M, Luche E, Gres S, Baylac A, Bergé M, Cenac C, Waget A, Klopp P, Iacovoni J, Klopp C, Mariette J, Bouchez O, Lluch J, Ouarné F, Monsan P, Valet P, Roques C, Amar J, Bouloumié A, Théodorou V, Burcelin R: Metabolic adaptation to a high-fat diet is associated with a change in the gut microbiota. Gut 2012;61:543–553

35 Burcelin R, Crivelli V, Dacosta A, Roy-Tirelli A, Thorens B: Heterogeneous metabolic adaptation of C57BL/6J mice to high-fat diet. Am J Physiol Endocrinol Metab 2002;282:E834–E842.

36 Membrez M, Blancher F, Jaquet M, Bibiloni R, Cani PD, Burcelin RG, Corthesy I, Macé K, Chou CJ: Gut microbiota modulation with norfloxacin and ampicillin enhances glucose tolerance in mice. FASEB J 2008;22:2416–2426.

37 Larsen N, Vogensen FK, van den Berg FWJ, Nielsen DS, Andreasen AS, Pedersen BK, Al-Soud WA, Sørensen SJ, Hansen LH, Jakobsen M: Gut microbiota in human adults with type 2 diabetes differs from non-diabetic adults. PLoS One 2010;5:e9085.

38 Wu X, Ma C, Han L, Nawaz M, Gao F, Zhang X, Yu P, Zhao C, Li L, Zhou A, Wang J, Moore JE, Millar BC, Xu J: Molecular characterisation of the faecal microbiota in patients with type II diabetes. Curr Microbiol 2010;61:69–78.

39 Furet JP, Kong LC, Tap J, Poitou C, Basdevant A, Bouillot JL, Mariat D, Corthier G, Doré J, Henegar C, Rizkalla S, Clément K: Differential adaptation of human gut microbiota to bariatric surgery-induced weight loss. Diabetes 2010;59:3049–3057.

40 Qin J, Li Y, Cai Z, Li S, Zhu J, Zhang F, Liang S, Zhang W, Guan Y, Shen D, Peng Y, Zhang D, Jie Z, Wu W, Qin Y, Xue W, Li J, Han L, Lu D, Wu P, Dai Y, Sun X, Li Z, Tang A, Zhong S, Li X, Chen W, Xu R, Wang M, Feng Q, Gong M, Yu J, Zhang Y, Zhang M, Hansen T, Sanchez G, Raes J, Falony G, Okuda S, Almeida M, LeChatelier E, Renault P, Pons N, Batto JM, Zhang Z, Chen H, Yang R, Zheng W, Li S, Yang H, Wang J, Ehrlich SD, Nielsen R, Pedersen O, Kristiansen K, Wang J: A metagenome-wide association study of gut microbiota in type 2 diabetes. Nature 2012;490:55–60.

41 Vrieze A, Van Nood E, Holleman F, Salojärvi J, Kootte RS, Bartelsman JF, Dallinga- Thie GM, Ackermans MT, Serlie MJ, Oozeer R, Derrien M, Druesne A, Van Hylckama Vlieg JE, Bloks VW, Groen AK, Heilig HG, Zoetendal EG, Stroes ES, de Vos WM, Hoekstra JB, Nieuwdorp M: Transfer of intestinal microbiota from lean donors increases insulin sensitivity in subjects with metabolic syndrome. Gastroenterology 2012;143: 913–916.

42 Wang Z, Klipfell E, Bennett BJ, Koeth R, Levison BS, DuGar B, Feldstein AE, Britt EB, Fu X, Chung YM, Wu Y, Schauer P, Smith JD, Allayee H, Tang WH, DiDonato JA, Lusis AJ, Hazen SL: Gut flora metabolism of phosphatidylcholine promotes cardiovascular disease. Nature 2011;472:57–63.

43 Wiedermann CJ, Kiechl S, Dunzendorfer S, Schratzberger P, Egger G, Oberhollenzer F, Willeit J: Association of endotoxemia with carotid atherosclerosis and cardiovascular disease: prospective results from the Bruneck study. J Am Coll Cardiol 1999;34:1975–1981.

44 Koren O, Spor A, Felin J, Fåk F, Stombaugh J, Tremaroli V, Behre CJ, Knight R, Fagerberg B, Ley RE, Bäckhed F: Human oral, gut, and plaque microbiota in patients with atherosclerosis. Proc Natl Acad Sci USA 2011;108:4592–4598

45 Miele L, Valenza V, La Torre G, Montalto M, Cammarota G, Ricci R, Mascianà R, Forgione A, Gabrieli ML, Perotti G, Vecchio FM, Rapaccini G, Gasbarrini G, Day CP, Grieco A: Increased intestinal permeability and tight junction alterations in nonalcoholic fatty liver disease. Hepatology 2009;49:1877–1887.

46 Dumas ME, Barton RH, Toye A, Cloarec O, Blancher C, Rothwell A, Fearnside J, Tatoud R, Blanc V, Lindon JC, Mitchell SC, Holmes E, McCarthy MI, Scott J, Gauguier D, Nicholson JK: Metabolic profiling reveals a contribution of gut microbiota to fatty liver phenotype in insulin-resistant mice. Proc Natl Acad Sci USA 2006;103:12511–12516.

47 Harte AL, Da Silva NF, Creely SJ, McGee KC, Billyard T, Youssef-Elabd EM, Tripathi G, Ashour E, Abdalla MS, Sharada HM, Amin AI, Burt AD, Kumar S, Day CP, McTernan PG: Elevated endotoxin levels in non-alcoholic fatty liver disease. J Inflamm 2010;7:15.

48 Henao-Mejia J, Elinav E, Jin C, Hao L, Mehal WZ, Strowig T, Thaiss CA, Kau AL, Eisenbarth SC, Jurczak MJ, Camporez JP, Shulman GI, Gordon JI, Hoffman HM, Flavell RA: Inflammasome-mediated dysbiosis regulates progression of NAFLD and obesity. Nature 2012;482:179–185.

49 Murphy EF, Cotter PD, Hogan A, O'Sullivan O, Joyce A, Fouhy F, et al: Divergent metabolic outcomes arising from targeted manipulation of the gut microbiota in diet-induced obesity. Gut 2012, E-pub ahead of print.

50 Kang JH, Yun SI, Park HO: Effects of Lactobacillus gasseri BNR17 on body weight and adipose tissue mass in diet-induced overweight rats. J Microbiol 2010;48:712–714.

51 Aronsson L, Huang Y, Parini P, Korach-André M, Håkansson J, Gustafsson JÅ, Pettersson S, Velmurugeson A, Rafter J: Decreased fat storage by Lactobacillus paracasei is associated with increased levels of angiopoietin-like 4 protein (ANGPTL4). PLoS One 2010;5:e13087.

52 Tanida M, Shen J, Maeda K, Horii Y, Yamano T, Fukushima Y, Nagai K: High-fat diet-induced obesity is attenuated by probiotic strain Lactobacillus paracasei ST11 (NCC2461) in rats. Obes Res Clin Pract 2008;2:159–169.

53 Lee K, Paek K, Lee H, Park JH, Lee Y: Antiobesity effect of trans-10, cis-12-conjugated linoleic acid-producing Lactobacillus plantarum PL62 on diet-induced obese mice. J Appl Micro 2007;103:1140–1146.

54 Lee HY, Park JH, Seok SH, Baek MW, Kim DJ, Lee KE, Paek KS, Lee Y, Park JH: Human originated bacteria, Lactobacillus rhamnosus PL60, produce conjugated linoleic acid and show anti-obesity effects in diet-induced obese mice. Biochim Biophy Acta 2006; 1761:736–744.

55 Takemura N, Okubo T, Sonoyama K: Lactobacillus plantarum strain No. 14 reduces adipocyte size in mice fed high-fat diet. Exp Biol Med 2010;235:849–856.

56 Ji Y, Kim H, Park H, Lee J, Yeo S, Yang J, Park SY, Yoon HS, Cho GS, Franz CMAP, Bomba A, Shin HK, Hilzapfel: Modulation of the murine microbiome with a concomitant anti-obesity effect by Lactobacillus rhamnosus GG and Lactobacillus sakei NR28. Benef Microbes 2012;3:13–22.

57 Kondo S, Xiao JZ, Satoh T, Odamaki T, Takahashi S, Sugahara H, Yaeshima T, Iwatsuki K, Kamei A, Abe K: Antiobesity effects of Bifidobacterium breve strain B-3 supplementation in a mouse model with high-fat diet-induced obesity. Biosci Biotech Biochem 2010; 74:1656–1661.

58 Xiao J, Kondo S, Takahashi N, Miyaji K, Oshida K, Hiramatsu A, Iwatsuki K, Kokubo S, Hosono A: Effects of milk products fermented by Bifidobacterium longum on blood lipids in rats and healthy adult male volunteers. J Dairy Sci 2003;86:2452–2461.

59 Hamad EM, Sato M, Uzu K, Yoshida T, Higashi S, Kawakami H, Kadooka Y, Matsuyama H, Abd El-Gawad IA, Imaizumi K: Milk fermented by Lactobacillus gasseri SBT2055 influences adipocyte size via inhibition of dietary fat absorption in Zucker rats. Br J Nutr 2009;101:716–724.

60 Yadav H, Jain S, Sinha P: Antidiabetic effect of probiotic dahi containing Lactobacillus acidophilus and Lactobacillus casei in high fructose fed rats. Nutrition 2007;23:62–68.

61 Martin FPJ, Wang Y, Sprenger N, Yap IKS, Lundstedt T, Lek P, Rezzi S, Ramadan Z, van Blanderen P, Fay LB, Kochhar S, Lindon JC, Holmes E, Nicholson JK: Probiotic modulation of symbiotic gut microbial-host metabolic interactions in a humanized microbiome mouse model. Mol Syst Biol 2008;4:157.

62 Yin YN, Yu QF, Fu N, Liu XW, Lu FG: Effects of four bifidobacteria on obesity in high-fat diet induced rats. World J Gastroenterol 2010;16:3394–3401.

63 Sato M, Uzu K, Yoshida T, Hamad EM, Kawakami H, Matsuyama H, Abd El-Gawad IA, Imaizuma K: Effects of milk fermented by Lactobacillus gasseri SBT2055 on adipocyte size in rats. Br J Nutr 2008;99: 1013–1017.

64 Kadooka Y, Ogawa A, Ikuyama K, Sato M: The probiotic *Lactobacillus gasseri* SBT2055 inhibits enlargement of visceral adipocytes and upregulation of serum soluble adhesion molecule (sICAM-1) in rats. Int Dairy J 2011;21:623–627.

65 Khan M, Raoult D, Richet H, Lepidi H, La Scola B: Growth-promoting effects of single-dose intragastrically administered probiotics in chickens. Br Poult Sci 2007;48:732–735.

66 Raoult D: Probiotics and obesity: a link? Nat Rev Micro 2009;7:616–616.

67 Delzenne N, Reid G: No causal link between obesity and probiotics. Nat Rev Microbiol 2009;7:901.

68 Wall R, Ross RP, Shanahan F, O'Mahony L, O'Mahony C, Coakley M, Hart O, Lawlor P, Quigley EM, Kiely B, Fitzgerald GF, Stanton C: Metabolic activity of the enteric microbiota influences the fatty acid composition of murine and porcine liver and adipose tissues. Am J Clin Nutr 2009;89:1393–1401.

69 Yadav H, Jain S, Sinha PR: Oral administration of dahi containing probiotic *Lactobacillus acidophilus* and *Lactobacillus casei* delayed the progression of streptozotocin-induced diabetes in rats. J Dairy Res 2008;75:189–195.

70 Tabuchi M, Ozaki M, Tamura A, Yamada N, Ishida T, Hosoda M, Hosono A: Antidiabetic effect of *Lactobacillus* GG in streptozotocin-induced diabetic rats. Biosci Biotech Biochem 2003;67:1421–1424.

71 Ma X, Hua J, Li Z: Probiotics improve high fat diet-induced hepatic steatosis and insulin resistance by increasing hepatic NKT cells. J Hepatol 2008;49:821–830.

72 Naito E, Yoshida Y, Makino K, Kounoshi Y, Kunihiro S, Takahashi R, Matsuzaki T, Miyazaki K, Ishikawa F: Beneficial effect of oral administration of *Lactobacillus casei* strain Shirota on insulin resistance in diet-induced obesity mice. J Applied Microbiol 2011;110:650–657.

73 Kadooka Y, Sato M, Imaizumi K, Ogawa A, Ikuyama K, Akai Y, Okano M, Kagoshima M, Tsuchida T: Regulation of abdominal adiposity by probiotics (*Lactobacillus gasseri* SBT2055) in adults with obese tendencies in a randomized controlled trial. Eur J Clin Nutr 2010;64:636–643.

74 Laitinen K, Poussa T, Isolauri E: Probiotics and dietary counselling contribute to glucose regulation during and after pregnancy: a randomised controlled trial. Br J Nutr 2009;101:1679–1687.

75 Andreasen AS, Larsen N, Pedersen-Skovsgaard T, Berg RMG, Moller K, Svendsen KD, Jakobsen M, Pedersen BK: Effects of *Lactobacillus acidophilus* NCFM on insulin sensitivity and the systemic inflammatory response in human subjects. Br J Nutr 2010;104:1831–1838.

76 Bukowska H, Pieczul-Mroz J, Jastrzebska M, Chełstowski K, Naruszewicz M: Decrease in fibrinogen and LDL-cholesterol levels upon supplementation of diet with *Lactobacillus plantarum* in subjects with moderately elevated cholesterol. Atherosclerosis 1998;137:437–438.

77 Naruszewicz M, Johansson ML, Zapolska-Downar D, Bukowska H: Effect of *Lactobacillus plantarum* 299v on cardiovascular disease risk factors in smokers. Am J Clin Nutr 2002;76:1249–1255.

78 Kullisaar T, Songisepp E, Mikelsaar M, Zilmer K, Vihalemm T, Zilmer M: Antioxidative probiotic fermented goats' milk decreases oxidative stress-mediated atherogenicity in human subjects. Br J Nutr 2003;90:449–456.

79 Kiessling G, Schneider J, Jahreis G: Long-term consumption of fermented dairy products over 6 months increases HDL cholesterol. Eur J Clin Nutr 2002;56:843–849.

80 Lewis S, Burmeister S: A double-blind placebo-controlled study of the effects of *Lactobacillus acidophilus* on plasma lipids. Eur J Clin Nutr 2005;59:776–780.

81 Simons LA, Amansec SG, Conway P: Effect of *Lactobacillus fermentum* on serum lipids in subjects with elevated serum cholesterol. Nutr Metab Cardiovasc 2006;16:531–535.

82 Greany K, Bonorden M, Hamilton-Reeves J, McMullen M, Wangen K, Phipps W, Feirtag J, Thomas W, Kurzer MS: Probiotic capsules do not lower plasma lipids in young women and men. Eur J Clin Nutr 2007;62:232–237.

Dr. Eileen F. Murphy
Alimentary Health Ltd.
Cork (Ireland)
E-Mail emurphy@ahealth.ie

Guarino A, Quigley EMM, Walker WA (eds): Probiotic Bacteria and Their Effect on Human Health and Well-Being.
World Rev Nutr Diet. Basel, Karger, 2013, vol 107, pp 122–127 (DOI: 10.1159/000345739)

Neonatal Necrotizing Enterocolitis

Josef Neu

Neonatal Biochemical Nutrition and GI Development Laboratory, Department of Pediatrics, Division of
Neonatology, University of Florida, Gainesville, Fla., USA

Abstract

Necrotizing enterocolitis (NEC) is a severe gastrointestinal disease that primarily affects preterm infants. The clinical presentation of this disease is variable, but often progresses rapidly to severe enterocolitis and death. The pathophysiology has been difficult to delineate, in part because it is likely to be multifactorial. Good biomarkers are not available. Here we address some of the clinical and laboratory manifestations, pathophysiology, and new research areas pertaining to NEC.

Copyright © 2013 S. Karger AG, Basel

Clinical Presentation

Necrotizing enterocolitis (NEC) is the most common severe neonatal gastrointestinal emergency that predominantly affects premature infants. The mortality of NEC ranges between 20 and 30%, with the greatest mortality among those requiring surgery [1]. The excessive and dysregulated inflammatory response initiated in the highly immunoreactive intestine in NEC spreads systemically and disrupts distant organs such as the brain and the retina, increasing the risk of neurodevelopmental delays and retinopathy of prematurity [2, 3]. Isolated or focal intestinal perforation, another entity that is sometimes confused with NEC, is not generally accompanied by an inflammatory component or by diffuse necrosis. This disease occurs earlier than NEC and is strongly associated with the combined use of glucocorticoids and early postnatal use of indomethacin [4, 5].

In the classic form of NEC seen most commonly in preterm infants, there is an inverse relationship between the gestational age and prevalence rate of NEC. The smallest and the most premature infants are at greatest risk [6]. Symptoms may progress rapidly, often within hours, from initial subtle signs to abdominal discoloration, intestinal perforation, and peritonitis, leading to systemic hypotension requiring intensive medical and/or surgical support.

The pathognomonic findings on the abdominal radiograph are pneumatosis intestinalis (intramural gas) and/or portal venous gas. Early imaging signs that should raise the suspicion of NEC include dilated bowel loops, paucity of gas, and gas-filled loops of bowel that are unaltered on repeated examinations and referred to as 'fixed bowel loops', indicating adynamic bowel loops. Extraluminal or 'free air' outside the bowel is a sign of advanced NEC. Plain abdominal radiography is the current modality of choice for the evaluation of neonates suspected of having NEC. However, abdominal sonography may add information on bowel wall thickness, echogenicity, peristalsis, and perfusion that may affect management [7, 8] and could become an important adjunct in detecting intramural gas and portal venous gas [7, 9–11].

Laboratory Features

Laboratory tests (platelets, C-reactive protein, white blood cells) and imaging techniques are currently used to confirm the diagnosis of NEC; however, these are suboptimal. A recent study showed that the urinary markers I-FABP and claudin-3, indicating loss of integrity of intestinal barrier, and the fecal marker calprotectin, representing intestinal inflammation, were significantly increased in neonates suspected of NEC who later developed NEC compared with neonates with other diagnoses, making these new noninvasive markers promising for the early diagnosis of NEC [12]. However, screening for NEC before clinical suspicion with such methodologies does not seem useful because I-FABP levels were not elevated prior to clinical suspicion of NEC, supporting the hypothesis that NEC is a disease with an acute onset with rapid clinical deterioration. Predictive biomarkers are clearly needed.

Pathogenesis

The pathophysiology of 'classic NEC' is incompletely understood. However, epidemiologic observations strongly suggest a multifactorial etiology (fig. 1) [6, 13]. The concurrence of a genetic predisposition, intestinal immaturity, and imbalanced microvascular tone, accompanied by a strong likelihood for abnormal microbial intestinal colonization and a highly dysregulated immunoreactive intestinal mucosa, present a confluence of predisposing factors. These have been discussed in more detail in other publications [6, 13].

Microbial Colonization
One hypothesis that is pertinent to the rest of this review involves the intestinal microbial ecology. Although specific pathogens have been cultured in outbreaks of NEC in single institutions, no organism has consistently been implicated. The human microbiome project initiated in 2007 [14] in conjunction with technological advances that allow for the molecular identification of a vast array of microbes that are difficult or impossible to culture from the intestine should provide new clues about the patho-

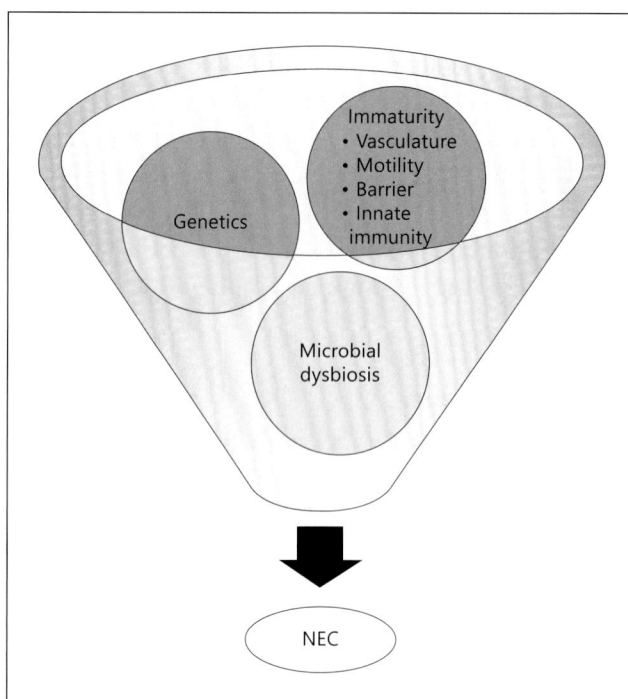

Fig. 1. Multifactorial patho-physiology of NEC. Several factors contribute to the classic form of the disease including genetics, an immature intestine, and environment.

genesis. Preliminary studies utilizing molecular methods to evaluate fecal microbiota from unaffected preterm infants, as well as some who developed NEC and had samples prior to and during NEC [15, 16], suggest an association with both unusual intestinal microbial species and overall less diverse microbiota, especially when there has been previous prolonged antibiotic therapy.

Prevention

Table 1 shows several strategies that have been studied for the prevention of NEC. The use of human milk and careful advancement of feedings appear to provide the greatest benefit, while the others either have not undergone adequate study or still are unproven.

Human Milk
Providing human milk to premature infants has been shown to decrease the incidence of NEC [17, 18]. An immature enteric nervous system and intestinal dysmotility warrants gradual and cautious increments in enteral feedings. Aggressive and improper enteral feeding is a risk factor associated with NEC [19, 20]. This has, in many cases, caused neonatologists to institute an overly cautious approach in which they withhold enteral feedings for prolonged periods followed by an equally inappropriate approach

Table 1. Prevention of NEC

Preventative measures
Current measures
Human milk
Careful advancement of feedings
Questionable measures
Formula acidification
IgG-IgA
Oral antibiotics
Glucocorticoids
Anticytokine therapy
Pre-, pro-, and postbiotics

Human milk and careful progression of feedings are widely considered to be beneficial, whereas the other factors have either been disproven or require additional study.

with aggressive feeding. Prolonged periods of nulla per os (NPO) are known to cause atrophy of the intestinal mucosa and result in overall delayed development of absorptive function, motility, and exocrine hormone secretion, and shifts the intestinal inflammatory response to one that favors the proinflammatory cytokines and chemokines over the anti-inflammatory mediators [21]. Initiating low-volume feedings appropriate for gestational age and increasing it gradually is a safe approach that would not be contraindicated while the infant is on ventilator support, using umbilical catheters, or receiving drugs such as indomethacin or dopamine [22].

Probiotics
Several trials over the past decade have evaluated the effects of various probiotics to prevent NEC [23, 24]. Meta-analyses across the available randomized trials have prompted recent commentaries suggesting the routine use of probiotics [25]. However, caution in the adoption of probiotics has been advised [25].

Microbial Components That Modulate Inflammation
Studies in epithelial cells and in an infant-formula-fed rodent model suggest that dead microbes may be as effective as live microbes in modulating excessive inflammatory stimuli [26–28]. This is supported by a study of preterms that showed a decreased incidence of NEC with inactivated probiotics [29].

Future Directions
Because of the fulminant nature of NEC, it is unlikely that new treatment strategies will provide major breakthroughs in alleviating its mortality and morbidity. Preventive approaches are more likely to yield the best results. Currently, it is reasonable to assume that interventions aimed at the more proximal events offer the greatest likeli-

hood for effective prophylaxis when compared to interventions aimed at more distal components of the cascade. Clear diagnostic criteria including the development of highly sensitive and specific predictive and diagnostic biomarkers [30] and new technologies that would detect predispositions to NEC need to be developed.

Acknowledgement

Supported in part by NIH RO1 HD 38954 to J.N.

References

1 Fitzgibbons SC, Ching Y, Yu D, et al: Mortality of necrotizing enterocolitis expressed by birth weight categories. J Pediatr Surg 2009;44:1072–1076.

2 Lodha A, Asztalos E, Moore AM: Cytokine levels in neonatal necrotizing enterocolitis and long-term growth and neurodevelopment. Acta Paediatr 2010; 99:338–343.

3 Martin CR, Dammann O, Allred E, Patel S, O'Shea TM, Kuban KC, Leviton A: Neurodevelopment of extremely preterm infants who had necrotizing enterocolitis with or without late bacteremia. J Pediatr 2010;157:751–756.e751.

4 Stark AR, Carlo WA, Tyson JE, Papile LA, Wright LL, Shankaran S, Donovan EF, Oh W, Bauer CR, Saha S, Poole WK, Stoll BJ, Fanaroff AA, Ehrenkranz RA, Korones SB, Stevenson DK: Adverse effects of early dexamethasone treatment in extremely-low-birth-weight infants. N Engl J Med 2001;344:95–101.

5 Attridge JT, Clark R, Walker MW, Gordon PV: New insights into spontaneous intestinal perforation using a national data set: (1) SIP is associated with early indomethacin exposure. J Perinatol 2006;26:93–99.

6 Lin PW, Stoll BJ: Necrotising enterocolitis. Lancet 2006;368:1271–1283.

7 Epelman M, Daneman A, Navarro OM, Morag I, Moore AM, Kim JH, Faingold R, Taylor G, Gerstle JT: Necrotizing enterocolitis: review of state-of-the-art imaging findings with pathologic correlation. Radiographics 2007;27:285–305.

8 Bohnhorst B, Muller S, Dordelmann M, Peter CS, Petersen C, Poets CF: Early feeding after necrotizing enterocolitis in preterm infants. J Pediatr 2003;143: 484–487.

9 Dordelmann M, Rau GA, Bartels D, Linke M, Derichs N, Behrens C, Bohnhorst B: Evaluation of portal venous gas detected by ultrasound examination for diagnosis of necrotising enterocolitis. Arch Dis Child Fetal Neonatal Ed 2009;94:F183–F187.

10 Dilli D, Oguz SS, Ulu HO, Dumanli H, Dilmen U: Sonographic findings in premature infants with necrotising enterocolitis. Arch Dis Child Fetal Neonatal Ed 2009;94:F232–F233.

11 Dilli D, Suna Oguz S, Erol R, Ozkan-Ulu H, Dumanli H, Dilmen U: Does abdominal sonography provide additional information over abdominal plain radiography for diagnosis of necrotising enterocolitis in neonates? Pediatr Surg Int 2011;27:321–327.

12 Thuijls G, Derikx JP, van Wijck K, Zimmermann LJ, Degraeuwe PL, Mulder TL, Van der Zee DC, Brouwers HA, Verhoeven BH, van Heurn LW, Kramer BW, Buurman WA, Heineman E: Non-invasive markers for early diagnosis and determination of the severity of necrotizing enterocolitis. Ann Surg 2010;251: 1174–1180.

13 Neu J, Walker WA: Necrotizing enterocolitis. N Engl J Med 2011;364:255–264.

14 Turnbaugh PJ, Ley RE, Hamady M, Fraser-Liggett CM, Knight R, Gordon JI: The human microbiome project. Nature 2007;449:804–810.

15 Wang Y, Hoenig JD, Malin KJ, Qamar S, Petrof EO, Sun J, Antonopoulos DA, Chang E, Claud EC: 16S rRNA gene-based analysis of fecal microbiota from preterm infants with and without necrotizing enterocolitis. ISME J 2009;3:944–954.

16 Mshvildadze M, Neu J, Schuster J, Theriaque D, Li N, Mai V: Intestinal microbial ecology in premature infants assessed with non-culture-based techniques. J Pediatr 2010;156:20–25.

17 Lucas A, Cole TJ: Breast milk and neonatal necrotising enterocolitis. Lancet 1990;336:1519–1523.

18 McGuire W, Anthony MY: Donor human milk versus formula for preventing necrotising enterocolitis in preterm infants: systematic review. Arch Dis Child Fetal Neonatal Ed 2003;88:F11–F14.

19 Anderson DM, Kliegman RM: The relationship of neonatal alimentation practices to the occurrence of endemic necrotizing enterocolitis. Am J Perinatol 1991;8:62–67.

20 Berseth CL, Bisquera JA, Paje VU: Prolonging small feeding volumes early in life decreases the incidence of necrotizing enterocolitis in very low birth weight infants. Pediatrics 2003;111:529–534.

21 Kudsk KA: Current aspects of mucosal immunology and its influence by nutrition. Am J Surg 2002;183: 390–398.

22 Neu J, Zhang L: Feeding intolerance in very-low-birthweight infants: what is it and what can we do about it? Acta Paediatr Suppl 2005;94:93–99.

23 Deshpande G, Rao S, Patole S, Bulsara: Updated meta-analysis of probiotics for preventing necrotizing enterocolitis in preterm neonates. Pediatrics 2010; 125:921–930.

24 Alfaleh K, Bassler D: Probiotics for prevention of necrotizing enterocolitis in preterm infants. Cochrane Database Syst Rev 2008;23:CD005496.

25 Neu J: Routine probiotics for premature infants: let's be careful! J Pediatr 2011;158:672–674.

26 Zhang L, Li N, Caicedo R, Neu J: Alive and dead *Lactobacillus rhamnosus* GG decrease tumor necrosis factor-alpha-induced interleukin-8 production in Caco-2 cells. J Nutr 2005;135:1752–1756.

27 Lopez M, Li N, Kataria J, Russell M, Neu J: Live and ultraviolet-inactivated *Lactobacillus rhamnosus* GG decrease flagellin-induced interleukin-8 production in Caco-2 cells. J Nutr 2008;138:2264–2268.

28 Li N, Russell WM, Douglas-Escobar M, Hauser N, Lopez M, Neu J: Live and heat-killed *Lactobacillus rhamnosus* GG (LGG): effects on proinflammatory and anti-inflammatory cyto/chemokines in gastrostomy-fed infant rats. Pediatr Res 2009;66:203–207.

29 Awad H, Mokhtar H, Imam SS, Gad GI, Hafez H, Aboushady N: Comparison between killed and living probiotic usage versus placebo for the prevention of necrotizing enterocolitis and sepsis in neonates. Pak J Biol Sci 2010;13:253–262.

30 Young C, Sharma R, Handfield M, Mai V, Neu J: Biomarkers for infants at risk for necrotizing enterocolitis: clues to prevention? Pediatr Res 2009;65:91R–97R.

Josef Neu, MD
Neonatal Biochemical Nutrition and GI Development Laboratory
Department of Pediatrics, Division of Neonatology, University of Florida
Box J296, 1600 SW Archer Road, Room HD 513, Gainesville, FL 32610-0296 (USA)
E-Mail neuj@peds.ufl.edu

Guarino A, Quigley EMM, Walker WA (eds): Probiotic Bacteria and Their Effect on Human Health and Well-Being.
World Rev Nutr Diet. Basel, Karger, 2013, vol 107, pp 128–138 (DOI: 10.1159/000345741)

Role of Probiotics in Allergies

Roberto Berni Canani[a, b] · Margherita Di Costanzo[a] · Rita Nocerino[a] ·
Vincenza Pezzella[a] · Linda Cosenza[a] · Viviana Granata[a] ·
Ludovica Leone[a]

[a]Food Allergy Unit, Department of Translational Medicine – Pediatric Section, and [b]European Laboratory for
the Investigation of Food Induced Diseases, University of Naples 'Federico II', Naples, Italy

Abstract

Allergies continue to be a growing health concern for children living in Western countries. Over
the past few years several studies have assessed the relationship between alterations in gut microbiota and allergies in humans. It is known that the gut microbiota of atopic subjects is different in quality and quantity from that of healthy controls. These findings prompted the concept
that specific beneficial bacteria from the human intestinal microflora, designated probiotics, which
may restore the intestinal ecosystem could be considered as a potential alternative strategy for
the prevention and treatment of allergies with promising results.

Copyright © 2013 S. Karger AG, Basel

Alterations of commensal microflora are involved in various intestinal and extraintestinal diseases, including allergies. The human gut microbiota is a complex ecosystem that is estimated to be composed of approximately 10^{14} bacterial cells, which is
ten times more than the total number of human cells in the body. Approximately
400–500 bacterial species make up the gut microbiota [1]. The gut microbiome, a term
for the collective community of bacteria and their total genome capacity in the human
gut, is approximately 150 times larger than the human gene complement, with an estimated 3.3 million microbial genes, and has thus been referred to as the 'forgotten
organ' and a 'superorganism' [2–4].

Comprehensive knowledge of the entire gut microbiota is key to understanding
the relationships between the gut microbiota and diseases. The accurate analysis
of the complex bacterial community structure of gastrointestinal tract microbiota
has been difficult. Gut microbiota can be investigated either by traditional
bacteriological culture techniques, molecular techniques, or exploration of microbiota-associated metabolic characteristics. New techniques based on bacterial

RNA and DNA have been developed for investigating, identifying, and quantifying the intestinal microbiota. Fluorescent in situ hybridization (FISH) combined with flow cytometry is a high-throughput method based on 16S rRNA probe hybridization that reliably characterizes the composition of fecal microbiota [5] (table 1).

The relationship between alterations in gut microbiota and allergies has been assessed in humans [6]. First, it is known that the gut microbiota of atopic subjects is different from that of healthy controls [7]: the atopic condition is associated with increased levels of aerobic bacteria and low levels of lactobacilli [8]. Additionally, in the fecal microbiota of allergic infants, higher levels of Firmicutes and reduced levels of Bacteroidetes were found [9, 10].

Finally, it was demonstrated that bifidobacteria in infants suffering from atopic dermatitis induces a higher secretion of proinflammatory cytokines, while those of healthy infants induce a higher secretion of anti-inflammatory cytokines [11]. Starting from these data, modifications to intestinal microflora in allergic individuals through the use of food supplements such as probiotics have been proposed as a strategy to treat or prevent allergic diseases.

Probiotic research and industry have continued to grow from early observations, and the global sales of probiotic ingredients, supplements, and foods amounted to USD 21.6 billion in 2010 and are expected to reach USD 31.1 billion by 2015, with an annual growth rate of 7.6% for the next 5-year period [12]. However, the growth of the probiotic industry has not been paralleled by advances in basic probiotic research and clinical trials determining their efficacy. We have only begun to understand the mechanisms and limitations of probiotics, which vary greatly among strains and treated individuals [13]. Over the past few years, several studies have looked for a role for probiotics in the prevention and treatment of allergies, in particular in the field of food allergies (FAs), with promising results.

Probiotics for Food Allergy

FA continues to be a growing health concern for infants living in Western countries [14]. The long-term prognosis for the majority of affected infants is good, with 80–90% naturally acquiring tolerance by the age of 5 years. However, recent studies suggest that the natural history of FA is changing, with an increasing persistence until later ages. The pathogenetic mechanisms of oral tolerance are complex and not completely known, although numerous studies implicate gut-associated immunity and enteric microflora, and it has been suggested that an altered composition of intestinal microflora results in an unbalanced local and systemic immune response to food allergens [6]. It is hoped that a number of ongoing studies will help provide clearer recommendations concerning the use of probiotics and the role of early introduction of allergenic foods for the promotion of tolerance.

Table 1. Techniques used to characterize the gut microbiota (modified from [5])

Culture: isolation of bacteria on selective media	Advantages: cheap, semiquantitative Disadvantages: labor intensive, 30% of gut microbiota have been cultured to date
qPCR: amplification and quantification of 16S rRNA; reaction mixture contains a compound that fluoresces when it binds to double-stranded DNA	Advantages: phylogenetic identification, quantitative, fast Disadvantages: PCR bias, unable to identify unknown species
DGGE/TGGE: gel separation of 16S rRNA amplicons using denaturant/temperature	Advantages: fast, semiquantitative, bands can be excised for further analysis Disadvantages: no phylogenetic identification, PCR bias
T-RFLP: fluorescently labeled primers are amplified and restriction enzymes are used to digest the 16S rRNA amplicon; digested fragments separated by gel electrophoresis	Advantages: fast, semiquantitative, cheap Disadvantages: no phylogenetic identification, PCR bias, low resolution
FISH: Fluorescently labeled oligonucleotide probes hybridize complementary target 16S rRNA sequences; fluorescence can be enumerated using flow cytometry	Advantages: phylogenetic identification, semiquantitative, no PCR bias Disadvantages: dependent on probe sequences, unable to identify unknown species
DNA microarrays: fluorescently labeled oligonucleotide probes hybridize with complementary nucleotide sequences; fluorescence detected with a laser	Advantages: phylogenetic identification, semiquantitative, fast Disadvantages: cross hybridization, PCR bias, species present in low levels can be difficult to detect
Cloned 16S rRNA gene sequencing: cloning of full-length 16S rRNA amplicon, Sanger sequencing and capillary electrophoresis	Advantages: phylogenetic identification, quantitative Disadvantages: PCR bias, laborious, expensive, cloning bias
Direct sequencing of 16S rRNA amplicons: massive parallel sequencing of partial 16S rRNA amplicons	Advantages: phylogenetic identification, quantitative, fast, identification of unknown bacteria Disadvantages: PCR bias, expensive, laborious
Microbiome shotgun sequencing: massive parallel sequencing of the whole genome	Advantages: phylogenetic identification, quantitative Disadvantages: expensive, analysis of data is computationally intense

DGGE = Denaturing gradient gel electrophoresis; qPCR = quantitative PCR; TGGE = temperature gradient gel electrophoresis; T-RFLP = terminal restriction fragment length polymorphism.

Oral Tolerance and Intestinal Microflora

Food antigens and intestinal microflora constitute the majority of the antigen load in the intestine, and the 'default' reaction of the immune system confronted with them leads to systemic unresponsiveness. This phenomenon is known as oral tolerance and is a key feature of intestinal immunity [15]. The complex interaction between intestinal contents and immune and nonimmune cells results in an environment that favors the tolerance by the induction of IgA antibodies and CD4+ T regulatory cells (producing IL-10 and IFN-γ) [16]. This ensures that a homeostatic balance is maintained between the intestinal immune system and its antigen load so that it retains the ability to recognize dangerous and harmless antigens as foreign and preserves the integrity of the intestinal mucosa. The inappropriate immune responses to food and intestinal microflora that are responsible for FA are a result of a deregulation of these crucial processes [17].

An allergic reaction mainly corresponds to the activation of Th2 cells against food allergens and occurs in two phases. The first phase corresponds to the transport of the allergen through the intestinal barrier, its capture by antigen-presenting cells (dendritic cells or enterocytes), and its presentation to naive Th0 cells, which differentiate in the presence of IL-4 into Th2 cells. Activated Th2 cells then produce an IL-4 cytokine that enables the production of allergen-specific IgE by B cells [18]. These secreted IgE antibodies then bind to mast cells via the IgE receptor FcERI. The activation phase corresponds to the degranulation of mast cells after further exposure to the same allergen that links directly with specific IgE on the surface of these cells. This phenomenon triggers release of allergic mediators involved in clinical manifestations of allergy (fig. 1).

In addition to acute allergic reactions triggered by IgE-mediated immune responses to food proteins, there are cell-mediated manifestations. The gut microbiota plays a crucial role in the establishment of tolerance to food antigens. Studies about axenic mice have revealed a failure in the acquisition of tolerance to food proteins [19]. Moreover, a transplanted healthy infant microbiota had a protective impact on sensitization and cow's milk allergy (CMA) in mice [20]. Thus, modifications to gut microbiota in allergic individuals through the use of food supplements such as probiotics could constitute an interesting strategy to treat or prevent FAs in the future.

From Animal Models to Human Studies

As we mentioned earlier, manipulation of the intestinal microflora has emerged as a potential alternative strategy for the prevention and treatment of FA. Probiotics, which may restore the intestinal ecosystem and affect the functioning of the gastrointestinal tract by a variety of mechanisms of action, have been proposed in this condition. Numerous animal and human studies have been performed to test the potential effects of various strains of probiotic bacteria. In this context, one of the most extensively studied probiotic has been *Lactobacillus* GG (LGG) [21].

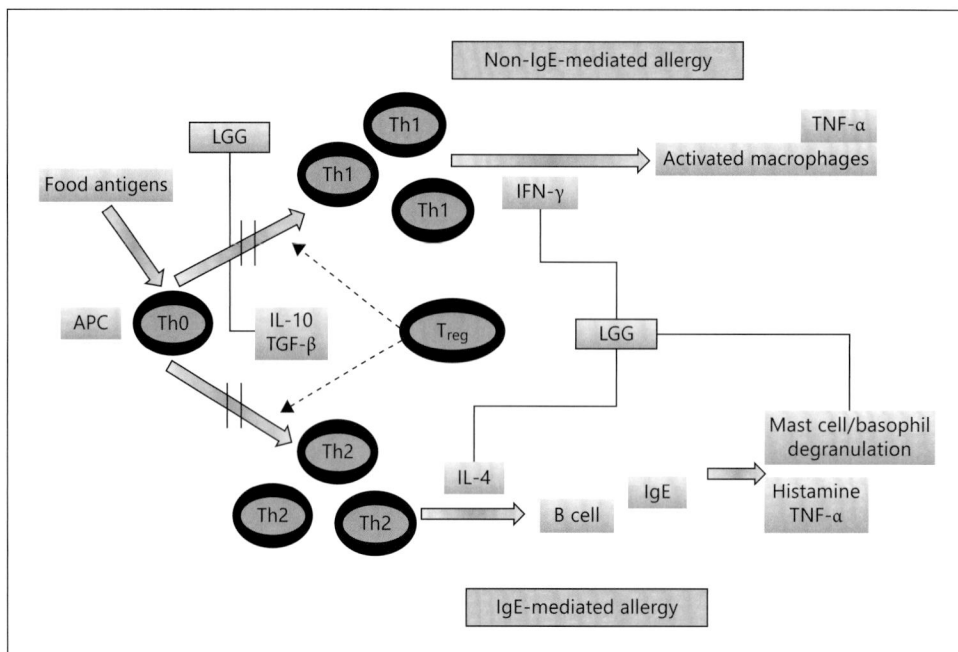

Fig. 1. LGG influences the mechanisms of FAs.

LGG is a bacterial strain isolated from human feces, capable of transiently colonizing the human intestine. Preventive and therapeutic properties of LGG related to atopic diseases, particularly in infants with CMA, have been reported. CMA is the most common FA in early childhood, with an estimated incidence ranging between 2 and 3% in infants and marginally lower in older children [22].

The potential mechanisms of action of LGG against FA are certainly multiple and exerted at different levels. Local influences of probiotics potentially include hydrolysis of antigenic peptides in the gut lumen, modulation of intestinal permeability and reduction of systemic penetration of antigens, increased local IgA production, modulation of local inflammation, and stimulation of epithelial cell growth and differentiation. Some possible systemic effects consist of anti-inflammatory effects mediated by Toll-like receptors, Th1 skewing of responses to allergens, and activation of tolerogenic dendritic cells, in addition to the production of T regulatory cells and tolerance acquisition [23–25].

Animal Studies

Animal models for FA provide an interesting tool to perform mechanistic research and to investigate the safety and efficacy of new therapeutic and preventive approaches for FA. Much progress has been made in recent years in developing an animal

Berni Canani · Di Costanzo · Nocerino · Pezzella · Cosenza · Granata · Leone model of CMA. In particular, animal models for CMA using oral sensitization mimick the human situation as children are most likely sensitized to cow's milk via the oral route.

Different mouse strains have been used to study CMA, showing that it can be induced in C3H/HeJ [26], C3H/HeOuJ [27], and BALB/c [28] mice on sensitization with whole cow's milk or β-lactoglobulin, which is one of the major whey allergens. In these models, cholera toxin is used as a mucosal adjuvant that may affect epithelial integrity and elicits both systemic and mucosal immune responses skewing toward Th2 responsiveness at the level of dendritic cells [29].

Oral tolerance to cow's milk protein has been studied in these models aiming to prevent both systemic and mucosal responses. Animal studies have evaluated the effectiveness of partial and/or extensive hypoallergenic formulas in preventing CMA in mouse models. Together with cellular in vitro assays, these animal models are also used for predicting allergenicity and the tolerogenic potential of hypoallergenic infant formulas [30].

In BALB/c mice that were sensitized with cow's milk protein via the systemic route, oral LGG supplementation favorably modulated immune reactions by shifting Th2-dominated trends toward Th1-dominated responses [31]. In these animal models, de Kivit et al. [32] evaluated the effects of a dietary supplementation with a symbiotic containing prebiotic galacto- and fructooligosaccharides and probiotic *Bifidobacterium breve* M-16V in mice that were sensitized orally with whey. They demonstrated that dietary intervention with this symbiotic increased galectin-9 expression by intestinal epithelial cells and serum galectin-9 levels in mice, and correlated with reduced acute allergic skin reaction and mast cell degranulation.

Human Studies

Allergy Prevention
With the hope of altering the intestinal microflora and thereby modulating immune responses during early life, it is rational to investigate the potential role of probiotics for preventing allergic diseases. Although most of the focus has been on postnatal microbial effects, there is emerging evidence of intrauterine effects and new evidence that these may be conferred by epigenetic effects on gene expression during critical periods of early development [33]. Most randomized controlled trials enrolled infants at high risk for developing allergy [34–38], whereas others also enrolled mothers who received probiotic supplementation during late pregnancy [39–43]. Despite several randomized controlled trials to assess the effects of probiotics in allergy prevention, there are still no definitive benefits or recommendations [44].

In a double-blinded randomized controlled trial, LGG or a placebo was given initially to 159 women during the final 4 weeks of pregnancy. If the infant was at high risk for atopic disease (atopic eczema, allergic rhinitis, or asthma), the treatment was continued for 6 months after birth in both the lactating woman and her infant [45]. A total of 132 mother-infant pairs were randomly assigned to receive either placebo or LGG and treated for 6 months while breastfeeding. The primary study endpoint was chronic recurrent atopic eczema in the infant. Atopic eczema was diagnosed in 46 of 132 (35%) of these study children by 2 years of age. The frequency of atopic eczema in the LGG-treated group was 15 of 64 (23%) versus 31 of 68 (46%) in the placebo group (RR 0.51; 95% CI: 0.32–0.84; $p < 0.01$). The number of mother-infant pairs required to be treated with LGG to prevent 1 case of chronic recurrent atopic eczema was 4.5. By 4 years of age, eczema occurred in 26% of the infants in the group treated with LGG, compared to 46% in the placebo group (RR 0.57; 95% CI: 0.33–0.97; $p < 0.01$). However, only 67% of the original study group was analyzed at the 4-year follow-up. These results support a preventive effect for giving a probiotic to mothers late in pregnancy and to both mothers and infants during the first 6 months of lactation for the prevention of atopic eczema in infants who are at risk for atopic diseases.

Prescott and Nowak-Węgrzyn [44] reported that the most recent meta-analysis concludes that there is some evidence that a probiotic or a synbiotic containing LGG may reduce the incidence of eczema in infants at high risk for allergic disease, but that there is no reproducible evidence for other probiotics. Furthermore, the effects of LGG are not seen in all studies [43]. There is also no evidence that any probiotics prevent other allergic conditions or sensitization. Prescott and colleagues also advised caution based on methodological concerns with some of the included studies. Wide variations in the methods, strains, and protocols contribute to significant heterogeneity and further complicate the analysis and interpretation. The effects of probiotics are variable and appear to depend on the strain, timing, method of administration, host, and other environmental factors [35, 42].

Allergy Treatment
Human studies about probiotics as treatment for allergy have revealed varying impacts: a beneficial effect of probiotics causing a decrease in severity of the Scoring Atopic Dermatitis index (SCORAD) levels in patients who received probiotic-supplemented formulas, a moderate effect, or no effect [46]. LGG is the strain that has been most studied. The studies with this strain suggested a therapeutic effect in patients with atopic eczema. Administration of the probiotic LGG to highly selected food-allergic patients (age 2 years, challenge-proven, and mild-to-moderate eczema) improved the eczema score significantly [47]. Studies in infants with eczema who received formulas supplemented with LGG have shown benefit in decreasing gastrointestinal symptoms [48]. For instance, after a challenge study in infants allergic to cow's milk proteins, fecal IgA levels were detected to be higher and TNF-α levels were lower in the LGG-treated group compared to the placebo [49].

Nermes et al. [50] investigated the interaction of LGG with skin and gut microbiota and humoral immunity in infants with atopic dermatitis. This study showed a statistically significant decrease in IgA- and IgM-secreting cells 1 month after starting an intervention with extensively hydrolyzed casein formula supplemented with LGG. This might indirectly indicate that LGG enhances gut barrier function and accelerates immunologic maturation in infants with atopic dermatitis. In particular, the finding of a significant increase in memory B cells in LGG-treated infants calls for larger studies. For clinical practice, these findings may indicate that specific probiotics are able to afford protection against gut intraluminal antigens and provide control of infections via an accelerated immunologic maturation. Moreover, LGG induced IFN-γ secretion in infants with CMA and in infants with IgE-associated dermatitis but, interestingly, not in infants with no CMA. This supports the view that the pattern of intestinal microflora may be aberrant in infants with an atopic predisposition, and the beneficial effects of probiotics are evident only in this group [51].

Recently in a prospective study, Berni Canani et al. [52] demonstrated that an extensively hydrolyzed casein formula containing LGG was able to stimulate a faster acquisition of clinical tolerance in infants with IgE or non-IgE-mediated CMA. The acquisition of oral tolerance was also confirmed by the negative response of skin-prick tests and atopy patch tests shown by all patients with a negative double-blind, placebo-controlled food challenges [52, 53]. Moreover, the addition of LGG to an extensively hydrolyzed casein formula significantly improved the recovery of the inflamed colonic mucosa compared to extensively casein hydrolyzed formula alone in infants with blood in the stools and presumptive cow's milk-allergic colitis, as indicated indirectly by greater decreases in fecal calprotectin and in the number of infants with persistence of occult blood in stools after 1 month [54].

All of these findings suggest a disrupting approach for infants affected by CMA, namely an 'active dietotherapy' able to effectively treat the symptoms of CMA and to reduce the time of clinical tolerance acquisition. Furthermore, CMA is often the first manifestation of the so-called 'atopic march', characterized by the appearance of other allergic manifestations in later years, such as asthma, atopic eczema, and allergic rhinitis. In this regard, it is important for future studies to also examine the long-term effects of the addition of probiotics to infant formulas on atopic march.

Conclusions

Numerous data suggest gut microbiota as a potential target to prevent or treat FA, and in particular CMA. These findings prompted the concept that specific beneficial bacteria from the human intestinal microflora, designated probiotics, could restore immune system homeostasis in children with FA. The potential mechanisms of actions are multiple, ranging from modulation of intestinal microflora composition, to a direct effect on intestinal mucosa structure and function, and on the local and systemic immune responses. LGG is the single probiotic with the largest body of in vitro and in vivo evidence on possible effects in pediatric allergic disorder, which is partially based on the number of studies using this probiotic. Larger trials and basic research are needed to confirm these findings, to better define the mechanisms of action, and to evaluate the potential factors influencing the response in allergic subjects.

References

1 Steinhoff U: Who controls the crowd? New findings and old questions about the intestinal microflora. Immunol Lett 2005;99:12–16.

2 Qin J, et al: A human gut microbial gene catalogue established by metagenomic sequencing. Nature 2010;464:59–65.

3 O'Hara AM, Shanahan F: The gut flora as a forgotten organ. EMBO Rep 2006;7:688–693.

4 Gill SR, et al: Metagenomic analysis of the human distal gut microbiome. Science 2006;312:1355–1359.

5 Fraher MH, O'Toole PW, Quigley EM: Techniques used to characterize the gut microbiota: a guide for the clinician. Nat Rev Gastroenterol Hepatol 2012;9: 312–322.

6 Gigante G, Tortora A, Ianiro G, Ojetti V, Purchiaroni F, Campanale M, Cesario V, Scarpellini E, Gasbarrini A: Role of gut microbiota in food tolerance and allergies. Dig Dis 2011;29:540–549.

7 Wills-Karp M, Santeliz J, Karp CL: The germless theory of allergic disease: revisiting the hygiene hypothesis. Nat Rev Immunol 2001;1:69–75.

8 Bjorksten B: The gut microbiota: a complex ecosystem. Clin Exp Allergy 2006;36:1215–1217.

9 Björkstén B, Sepp E, Julge K, Voor T, Mikelsaar M: Allergy development and the intestinal microflora during the first year of life. J Allergy Clin Immunol 2001;108:516–520.

10 Nakayama J, Kobayashi T, Tanaka S, Korenori Y, Tateyama A, Sakamoto N, Kiyohara C, Shirakawa T, Sonomoto K: Aberrant structures of fecal bacterial community in allergic infants profiled by 16S rRNA gene pyrosequencing. FEMS Immunol Med Microbiol 2011;63:397–406.

11 He F, Morita H, Hashimoto H, Hosoda M, Kurisaki J, Ouwehand AC, Isolauri E, Benno Y, Salminen S: Intestinal Bifidobacterium species induce varying cytokine production. J Allergy Clin Immunol 2002; 109:1035–1036.

12 Agheyisi R: The probiotics market: ingredients, supplements, foods. BCC Research. Report: FOD013D. July 2011. http://www.bccresearch.com/report/nutraceuticals-markets-processing-technologies-fod013d.html.

13 Yan F, Polk DB: Probiotics: progress toward novel therapies for intestinal diseases. Curr Opin Gastroenterol 2010;26:95–101.

14 Allen KJ, Koplin JJ: The epidemiology of IgE-mediated food allergy and anaphylaxis. Immunol Allergy Clin North Am 2012;32:35–50.

15 Mowat AM: Anatomical basis of tolerance and immunity to intestinal antigens. Nat Rev Immunol 2003;3:331–341.

16 Gourbeyre P, Denery S, Bodinier M: Probiotics, prebiotics, and synbiotics: impact on the gut immune system and allergic reactions. J Leukoc Biol 2011;89: 685–695.

17 Vighi G, Marcucci F, Sensi L, et al: Allergy and the gastrointestinal system. Clin Exp Immunol 2008; 153:3–6.

18 Romagnani S: Regulation of the development of type 2 T-helper cells in allergy. Curr Opin Immunol 1994; 6:838–846.

19 Bjorksten B: Environmental influences on the development of the immune system: consequences for disease outcome. Nestle Nutr Workshop Ser Pediatr Program 2008;61:243–254.

20 Rodriguez B, Prioult G, Hacini-Rachinel F, et al: Infant gut microbiota is protective against cow's milk allergy in mice despite immature ileal T-cell response. FEMS Microbiol Ecol 2011;79:192–202.

21 Berni Canani R, Di Costanzo M, Pezzella V, Cosenza L, Granata V, Terrin G, Nocerino R: The potential therapeutic efficacy of Lactobacillus GG in children with food allergies. Pharmaceuticals 2012;5:655–664.

22 Apps JR, Beattie RM: Cow's milk allergy in children. Cont Med Educ 2009;339:b2275.

23 Sherman PM, Ossa JC, Johnson-Henry K: Unraveling mechanisms of action of probiotics. Nutr Clin Pract 2009;24:10.

24 Yan F, Cao H, Cover TL, Washington MK, Polk DB: Soluble proteins produced by probiotic bacteria regulate intestinal epithelial cell survival and growth. Gastroenterology 2007;132:562–575.

25 Gourbeyre P, Denery S, Bodinier M: Probiotics, prebiotics, and synbiotics: impact on the gut immune system and allergic reactions. J Leukoc Biol 2011;89: 685–695.

26 Li XM, Schofield BH, Huang CK, et al: A murine model of IgE-mediated cow's milk hypersensitivity. J Allergy Clin Immunol 1999;103:206–214.

27 Frossard CP, Steidler L, Eigenmann PA: Oral administration of an IL-10-secreting *Lactococcus lactis* strain prevents food-induced IgE sensitization. J Allergy Clin Immunol 2007;119:952–959.

28 von der Weid T, Bulliard C, Fritsche R: Suppression of specific and bystander IgE responses in a mouse model of oral sensitization to beta-lactoglobulin. Int Arch Allergy Immunol 2001;125:307–315.

29 Lycke N: The mechanism of cholera toxin adjuvanticity. Res Immunol 1997;148:504–520.

30 Fritsché R: Animal models in food allergy: assessment of allergenicity and preventive activity of infant formulas. Toxicol Lett 2003;140–141:303–309.

31 Thang CL, Baurhoo B, Boye JI, Simpson BK, Zhao X: Effects of *Lactobacillus rhamnosus* GG supplementation on cow's milk allergy in a mouse model. Allergy Asthma Clin Immunol 2011;7:20.

32 de Kivit S, Saeland E, Kraneveld AD, et al: Galectin-9 induced by dietary synbiotics is involved in suppression of allergic symptoms in mice and humans. Allergy 2012;67:343–352.

33 Prescott S, Saffery R: The role of epigenetic dysregulation in the epidemic of allergic disease. Clin Epigenet 2011;2:223–232.

34 Lodinová-Zádníková R, Prokesová L, Kocourková I, et al: Prevention of allergy in infants of allergic mothers by probiotic *Escherichia coli*. Int Arch Allergy Immunol 2010;153:201–206.

35 Prescott SL, Wiltschut J, Taylor A, et al: Early markers of allergic disease in a primary prevention study using probiotics: 2.5-year follow-up phase. Allergy 2008;63:1481–1490.

36 Scalabrin DM, Johnston WH, Hoffman DR, et al: Growth and tolerance of healthy term infants receiving hydrolyzed infant formulas supplemented with *Lactobacillus rhamnosus* GG: randomized, double-blind, controlled trial. Clin Pediatr (Phila) 2009;48: 734–744.

37 Soh SE, Aw M, Gerez I, et al: Probiotic supplementation in the first 6 months of life in at risk Asian infants – effects on eczema and atopic sensitization at the age of 1 year. Clin Exp Allergy 2009;39:571–578.

38 Taylor AL, Dunstan JA, Prescott SL: Probiotic supplementation for the first 6 months of life fails to reduce the risk of atopic dermatitis and increases the risk of allergen sensitization in high-risk children: a randomized controlled trial. J Allergy Clin Immunol 2007;119:184–191.

39 Kalliomäki M, Salminen S, Arvilommi H, et al: Probiotics in primary prevention of atopic disease: a randomised placebo-controlled trial. Lancet 2001; 357:1076–1079.

40 Kim JY, Kwon JH, Ahn SH, et al: Effect of probiotic mix (*Bifidobacterium bifidum*, *Bifidobacterium lactis*, *Lactobacillus acidophilus*) in the primary prevention of eczema: a double-blind, randomized, placebo-controlled trial. Pediatr Allergy Immunol 2010; 21:e386–e393.

41 Niers L, Martín R, Rijkers G, et al: The effects of selected probiotic strains on the development of eczema (the PandA study). Allergy 2009;64:1349–1358.

42 Wickens K, Black PN, Stanley TV, et al: A differential effect of 2 probiotics in the prevention of eczema and atopy: a double-blind, randomized, placebo-controlled trial. J Allergy Clin Immunol 2008;122:788–794.

43 Kopp MV, Salfeld P: Probiotics and prevention of allergic disease. Curr Opin Clin Nutr Metab Care 2009;12:298–303.

44 Prescott S, Nowak-Węgrzyn A: Strategies to prevent or reduce allergic disease. Ann Nutr Metab 2011;59: 28–42.

45 Kalliomäki M, Salminen S, Poussa T, Arvilommi H, Isolauri E: Probiotics and prevention of atopic disease: 4-year follow-up of a randomised placebo-controlled trial. Lancet 2003;361:1869–1871.

46 Kalliomaki M, Antoine JM, Herz U, et al: Guidance for substantiating the evidence for beneficial effects of probiotics: prevention and management of allergic diseases by probiotics. J Nutr 2010;140:713S–721S.

47 Majamaa H, Isolauri E: Probiotics: A novel approach in the management of food allergy. J Allergy Clin Immunol 1997;99:179–185.

48 Isolauri E, Arvola T, Sutas Y, et al: Probiotics in the management of atopic eczema. Clin Exp Allergy 2000;30:1604–1610.

49 Isolauri E: Studies on Lactobacillus GG in food hypersensitivity disorders. Nutr Today Suppl 1996;31: 285–315.

50 Nermes M, Kantele JM, Atosuo TJ, et al: Interaction of orally administered *Lactobacillus rhamnosus* GG with skin and gut microbiota and humoral immunity in infants with atopic dermatitis. Clin Exp Allergy 2010;41:370–377.

51 Pohjavuori E, Viljanen M, Korpela R, et al: Lactobacillus GG effect in increasing IFN-γ production in infants with cow's milk allergy. J Allergy Clin Immunol 2004;114:131–136.

52 Berni Canani R, Nocerino R, Terrin G, et al: Effect of extensively hydrolyzed casein formula supplemented with Lactobacillus GG on tolerance acquisition in infants with cow's milk allergy: a randomized trial. J Allergy Clin Immunol 2012;129:580–582, 582.e1–e5.

53 Baldassarre ME, Laforgia N, Fanelli M, et al: Lactobacillus GG improves recovery in infants with blood in the stools and presumptive allergic colitis compared with extensively hydrolyzed formula alone. J Pediatr 2010;156:397–401.

54 Di Costanzo M, Berni Canani R: Effects of an extensively hydrolyzed casein formula containing Lactobacillus GG in children with cow's milk allergy. Functional Food Rev 2012;4:70–76.

Roberto Berni Canani, MD, PhD
Food Allergy Unit, Department of Translational Medicine – Pediatric Section
University of Naples 'Federico II'
Via S Pansini 5, IT–80131 Naples (Italy)
E-Mail berni@unina.it

Guarino A, Quigley EMM, Walker WA (eds): Probiotic Bacteria and Their Effect on Human Health and Well-Being.
World Rev Nutr Diet. Basel, Karger, 2013, vol 107, pp 139–150 (DOI: 10.1159/000345743)

Probiotics in Respiratory Infections

Eugenia Bruzzese · Andrea Lo Vecchio · Eliana Ruberto

Department of Translational Medical Science-Section of Pediatrics, University Federico II of Naples,
Naples, Italy

Abstract

Probiotics are an established therapeutic or preventive intervention for gastrointestinal diseases and their efficacy is supported by several mechanisms, including the modulation of the immune response. However, there is evidence that the administration of probiotics may also be effective in the prevention or treatment of respiratory infections. Trials have been performed in general populations of infants and children and in populations at increased risk of respiratory infections, such as those in day care centers, schools, or hospitals, and evidence of a reduced rate of infections has been obtained in all those settings. In addition, the efficacy of probiotics to prevent respiratory infections has also been investigated in children with clinical conditions at increased risk of respiratory infections, such as those with atopy or with chronic diseases, and preliminary evidence of efficacy has been found. In adults, similar data have been obtained, although the body of evidence and the quality of data does not reach that available in childhood. Finally, the efficacy of probiotics in populations at risk of very severe respiratory infections, such as children with cystic fibrosis and preterm infants, has been investigated with promising results. However, differences in definitions, conditions, strains, and study protocols do not allow conclusions and further data are needed. In conclusion, probiotics provide a novel exciting opportunity for prevention and treatment of respiratory infections that is highly appealing in conditions of increased risk, particularly in children.

Acute respiratory infections (ARI), including both upper respiratory tract infections (URTI) and lower respiratory infections (LRTI), are the most common illnesses during infancy worldwide. In developed countries, most children experience several respiratory infections (6–8 colds per year) in their first years of life. The common cold is also frequent in adults with an occurrence of 2–4 colds per year.

Options for treatment of URTI are limited as it is prevalently of viral origin and generally based on symptoms. Probiotics have been proposed as therapy both in general settings and in populations at increased risk of URTI, which include conditions related to the host or to logistical conditions.

Day care centers and schools are an ideal place for the transmission of ARI. Children attending day care centers are at 2–3 times greater risk for developing ARI than children at home [1]. Atopy also represents an at-risk condition for developing respiratory infections. The high rates of respiratory infections translate into an increase in both social and family costs. ARI are the main reasons for a medical consultation and for an overuse of both prescriptions for drugs and investigations. Finally, a sick child is a major cause of missed work days.

Respiratory infections, together with gastrointestinal infections, also are the main cause of nosocomial infections in children in developed countries. The incidence of nosocomial infections is high, ranging from 5 to 44% [2, 3]. Although generally mild, nosocomial infections prolong the hospital stay and significantly increase hospital expenses. The current procedures to reduce the risk of nosocomial infections such as vaccination programs, hygiene measures, and visitor screening have only been partially effective in reducing their incidence.

In the last 10 years, it is becoming clear that selected probiotics may modulate the immune and inflammatory response not only at the gastrointestinal level, but also at the respiratory level. Probiotics have been used as a treatment for a number of diseases, such as acute gastroenteritis, necrotizing enterocolitis (NEC) in newborns, antibiotic-related diarrhea, and urinary tract infections in young women, with variable efficacy according to the specific condition and the strain used.

The available evidence in the field of respiratory infections is increasing although the data relating to the use of probiotics in the treatment of ARI are currently limited and weak if compared to those available for their prevention. Former trials proposed the restoration of the throat microflora through α-streptococci supplementation in addition to standard treatment in patients prone to respiratory tract infections (including pharyngotonsillitis and otitis). This intervention, following the antibiotic therapy, seems to reduce the recurrence rate in susceptible subjects [4–6]. A regular ingestion of fermented milk products containing probiotics reduced the presence of potentially pathogenic bacteria in the upper respiratory tract [7].

In this review we describe the effects of probiotic administration in respiratory tract infections in both children and adults, mainly addressing their efficacy in prevention for which the vast majority and the most promising results are available. In addition, we specifically reviewed the efficacy of probiotic administration in different clinical conditions such as healthy subjects, subjects at increased risk to develop an ARI, and children at risk because of an underlying chronic condition.

Probiotics in the Prevention of Respiratory Tract Infections in Children

Probiotic supplementation is a promising strategy to prevent respiratory infections in children. A recent Cochrane review, including 14 trials, reported a significant reduction of URTI and a limitation of antibiotic prescription in people of all ages receiving different probiotic formulations [8]. Although the vast majority of studies were performed in children, the authors did not perform any subgroup analysis. This complicates the interpretation of the results.

For evaluating the efficacy of probiotics, it is crucial to defining the setting and the study population [9], as well as the disease target of prevention. Often respiratory diseases are poorly defined (e.g. URTI, without any further feature). Data on probiotics are available in healthy children, children at risk to develop an ARI because of an increased exposure (day care centers, schools), and children with an underlying clinical condition that predisposes to an ARI such as atopy or an inflammatory chronic disease.

In the last decade several clinical trials have been performed wih children, but the available data are still heterogeneous and sometimes controversial due to the probiotic preparation used, its dosage, and the clinical setting. Moreover, the different frequencies of respiratory infections in different countries may influence results.

Trials have been performed in healthy children using a number of bacterial strains in developing and developed countries and most of published trials have reported a reduction of the incidence of ARI in children treated compared to placebo groups (table 1). In a recent trial, infants receiving *Bifidobacterium animalis* subsp. *lactis* BB-12 (BB-12) via a novel slow-release pacifier experienced fewer respiratory infections than control infants [10]. There has also been a randomized double-blind controlled study that included healthy 6-month infants exclusively formula fed: the control group received a formula supplemented with the prebiotic galactooligosaccharides and the experimental group received formula with the same amount of galactooligosaccharides but containing *Lactobacillus fermentum* CECT5716. A reduction of both the incidence of URTI and recurrent respiratory infections was observed in the experimental group, whereas no significant difference was observed in the incidence of LRTI [11]. Different probiotic strains were also tested in older healthy children. Leyer et al. [12] compared three groups of healthy children aged 3–5 years: one group receiving *L. acidophilus* NCFM, a second group receiving *L. acidophilus* NCFM and *B. animalis* subsp. *lactis* Bi-07, and a third group as the control. The incidence and duration of several respiratory symptoms were reduced in both probiotic groups compared to placebo, with a more pronounced effect in the preparation with two strains than in the single strain preparation. The effect was very clear, especially in the area of rhinorrhea. In addition, a reduced use of antibiotic and fewer days of absence from school were found in the intervention groups, thus suggesting a potential prophylactic role of probiotics versus cold and influenza-like symptoms.

Other studies have focused on the beneficial role of probiotics in preventing respiratory infections in specific settings at risk of infections because of overcrowding and exposure to the same agents such as day care centers and hospitals (table 1).

Weizman et al. [13] examined probiotic effects in infants aged 4–10 months attending day care centers. Enrolled infants were randomized to a humanized cow's

Table 1. Evidence on the use of probiotics in healthy and at-risk pediatric populations

Probiotic strain	Population	Effect of treatment	Reference
Healthy pediatric populations *B. animalis* subsp. *lactis* BB-12	healthy infants	reduction in respiratory infections	Taipale et al. [10], 2011
L. fermentum CECT5716	healthy infants aged 6-months	reduction in the incidence of URTI and in recurrent respiratory infections, no significant difference in the incidence of LRTI	Maldonado et al. [11], 2012
L. acidophilus NCFM; *L. acidophilus* NCFM and *B. animalis* subsp. *lactis* Bi-07	healthy children aged 3–5 years	reduction in incidence and duration of several respiratory symptoms and antibiotic use and days absence	Leyer et al. [12], 2009
At-risk pediatric populations			
B. lactis BB-12 or *L. reuteri*	infants aged 4–10 months attending day care centers	no significant difference in the rate and duration of respiratory illness	Weizman et al. [13], 2005
L. casei (DN-114 001)	healthy children aged 3–6 years attending day care center/school	reduction in the incidence rate of common infectious diseases (significant for URTI, not significant difference for LRTI) and in the mean number of days of medication use	Merenstein et al. [14], 2010
LGG	healthy children aged 1–6 years attending day care centers	reduction in number of days of absence, a slight reduction in the incidence of respiratory infections and antibiotic treatment	Hatakka et al. [15], 2001
LGG	children attending day care centers	reduction in number of URTI	Hojsak et [] [16], []al.[1] [16], 2010
LGG	hospitalized children	reduction in the risk for respiratory tract infections	Hojsak et al. [17], 2010
LGG	children affected by CF	reduction in pulmonary exacerbations and hospital admissions	Bruzzese et al. [19], 2007
L. acidophilus and *L. bulgaricus* and *B. bifidum* and *S. thermophiles*	children affected by CF	reduction in the pulmonary exacerbation rate	Weiss et al. [20], 2010

milk formula supplemented with *B. lactis* BB-12, *L. reuteri,* or the same formula without probiotics. No significant differences in the rate and duration of respiratory illnesses were observed between groups.

In a double-blind, randomized, placebo controlled clinical trial, 638 healthy children aged between 3 and 6 years and attending a day care center/school were enrolled. The intervention group received a fermented dairy drink containing *L. casei* (DN-114 001) for 3 months. The incidence rate of common infectious diseases including URTI, LRTI, and acute gastroenteritis was lower in the probiotic group, with a significant reduction of URTI. However, no significant difference was specifically observed in the incidence of LRTI between the two groups, but the mean number of days of medication use was significantly lower in the probiotic group compared to the control group [14].

Similar results were observed in healthy children aged 1–6 years attending day care centers and treated with a probiotic milk [containing *Lactobacillus* GG (LGG)] during the winter season. A reduced absence from day care due to illnesses and a slightly lower incidence of respiratory infections together with a reduction of antibiotic treatment were observed in the experimental group [15].

The efficacy of LGG was tested in children attending day care centers in the Zagreb area [16]. Children receiving a fermented diary product with LGG showed a lower number of URTI and the authors concluded that the use of a fermented milk may be recommended as a valid measure for decreasing the risk of URTI in children attending day care centers.

Another promising field of application is the prevention of nosocomial infections. Recently, a trial with fermented milk supplemented with LGG was effective in significantly reducing the risk for respiratory tract infections in hospitalized children, namely a reduced risk for ARI lasting more than 3 days was also observed [17]. In addition, promising but inconsistent results were recently described for otitis media [18].

Not only the setting, but also the presence of an underlying disease, may be associated with an increased susceptibility to infections. Children with cystic fibrosis (CF) frequently suffer from pulmonary exacerbations and often require antibiotic treatment. A prospective, randomized, placebo-controlled, crossover study investigated the effects of LGG given for 6 months on pulmonary exacerbation in CF children [19]. A reduction in pulmonary exacerbations and hospital admissions was observed in the treatment group compared with the control group. The efficacy of probiotics in preventing exacerbations in CF was confirmed in another trial by Weiss et al. [20] in a population of CF children with mild-to-moderate lung disease (FEV_1 >40%) chronically infected with *Pseudomonas aeruginosa.* The tested probiotic was a mixture of bacteria (*L. acidophilus, L. bulgaricus, B. bifidum, Streptococcus thermophiles)* given for 6 months. The pulmonary exacerbation rate was significantly reduced compared to the previous 2 years and to the 6-month posttreatment period.

Based on the available evidence, the efficacy of probiotics in preventing ARI in children is promising, but not conclusive. Results are preliminary and sometimes conflicting. This may depend on factors such as the study population (e.g. age, location, health status, risk factors, other therapies), the intervention (e.g. specific strain, other supplementation, formulation, dosage, duration of assumption), and the outcome measures (e.g. incidence of disease, severity of disease, type of infection, change of behavior, costs, medication use). However, if data can be confirmed, the use of probiotics in preventing ARI might have a major impact on the frequency of the infectious diseases in the pediatric age and might be recommended as a public health intervention.

Prevention of Respiratory Infections in the Adult Population

ARI are also the most common reasons for medical consultation for adults in developed areas, with an incidence of 2–6 episodes per year per person. It is also a major cause of work-day loss and accounts for 75% of all antibiotic use in high-income countries [21]. Although the bulk of evidence is available for children, using probiotics to reduce the incidence of ARI has been proposed for adult and elderly people, too.

A recent Cochrane review reported a significant reduction of subjects with episodes of URTI and a limitation of antibiotics in people of all ages receiving different probiotic formulations. The review included three studies in adults of about 40 years of age and in elderly people [8]. Looking at the data from single studies, a daily supplementation of probiotics (mainly lactobacilli and bifidobacteria) showed a slight effect on ARI in adults, elderly, and selected at-risk populations (table 2). However, either a combined probiotic preparation (including *L. gasseri, B. longum, and B. bifidum*) or a symbiotic formulation (including *L. rhamnosus, L. plantarum, and B. lactis*) demonstrated, in randomized controlled trials, a good efficacy in shortening the duration of episodes by about 2 days and the severity of symptoms [22–24].

Probiotics may also be indirectly effective for ARI in adults. The use of selected probiotics has also been proposed to enhance vaccination response in a healthy population. In a recent randomized controlled trial, a 6-week supplementation of either *B. animalis* BB-12 or *L. casei* 431 significantly increased antigen-specific immune response in healthy individuals receiving an influenza vaccination. The effect was related to an increased adaptive immune response to vaccination as shown by the vaccine-specific plasma IgG and salivary IgA, and the number of subjects achieving seroconversion [25].

In a previous study, Davidson et al. [26] studied the effects of LGG on immune response by live attenuated influenza vaccine in normal healthy adults. Although subjects receiving probiotics did not achieve greater seroprotection after administration of vaccine for the H1N1 and B strains, there was a significant improvement in those subjects receiving LGG for the H3N2 strain.

Table 2. Evidence on the use of probiotics in healthy and at-risk adult population

Probiotic strain	Population	Effect of treatment	Reference
Healthy adult population			
L. rhamnosus GG L. acidophilus 145 S. thermophilus Bifidobacterium sp. B420	healthy adults	reduction of nasal colonization with pathogens	Gluck et al. [40], 2003
L. plantarum L. paracasei	healthy adults	reduction of incidence and duration of common cold episodes reduction of severity of symptoms	Berggren et al. [22], 2010
L. gasseri PA 16/8 B. longum SP 07/3 B. bifidum MF 20/5	healthy adults	no effect on the incidence of common cold, but significant effect on duration of episodes (reduction by almost 2 days)	de Vrese et al. [24], 2006
Synbiotic preparations containing L. plantarum L. rhamnosus B. lactis Lactoferrin and prebiotics (FOS, GOS)	healthy adults	reduction of episodes of URTI and severity of illness	Pregliasco et al. [23], 2008
B. animalis ssp. lactis BB-12 L. paracasei	healthy adults undergoing vaccination	increase of vaccine specific-plasma immunoglobulins and of the number of subjects achieving seroconversion.	Rizzardini et al. [25], 2012
L. rhamnosus GG	healthy adults undergoing vaccination	effect on seroprotection evident only for H3N2 influenza strain	Davidson et al. [26], 2011
L. delbrueckii subsp. bulgaricus S. thermophilus	healthy adults and elderly	decreased risk of common cold or influenza virus	Makino et al. [41], 2010
At-risk adult population			
L. paracasei	healthy free-living elderly	reduction of duration of URTI, specifically rhinopharyngitis	Guillemard et al. [27], 2010
L. rhamnosus GG	marathon runners during training	no effect on the incidence of respiratory tract infections	Kekkonen et al. [31], 2007
L. fermentum PCC	athletes	no significant effect on the incidence of URTI	West et al. [32], 2011
L. casei strain DN- 114001	militaries during intense training	no significant effect on the incidence of URTI	Tiollier et al. [33], 2007
L. casei Shirota	athletes	reduce the frequency of respiratory infections, no difference in symptoms severity; higher saliva IgA in treatment group compared with controls	Gleeson et al. [34], 2011
L. casei DN 114001	shift workers (stressed population)	reduction of common infections and prolongation of disease-free time	Guillemard et al. [35], 2010

Table 2. Continued

Probiotic strain	Population	Effect of treatment	Reference
L. rhamnosus GG	patients in ICU	delayed occurrence of respiratory tract infections and of *P. aeruginosa* infections; no difference in the incidence of VAP	Forestier et al. [28], 2008
Synbiotic preparations containing *Pediococcus pentosaceus*, *Lactococcus raffinolactis*, *L. paracasei* subsp. *paracasei, L. plantarum*	patients in ICU	no difference in VAP occurrence and overall mortality; slight reduction in length of stay in ICU	Knight et al. [42], 2009
Synbiotic preparations containing *P. pentosaceus*, *L. raffinolactis*, *L. paracasei* subspecies *paracasei, L. plantarum*	patients in ICU	reduced incidence of VAP no difference in mortality	Spindler-Vesel et al. [43], 2007
Synbiotic preparations containing *P. pentosaceus*, *L. raffinolactis*, *L. paracasei* subspecies *paracasei, L. plantarum*	patients in ICU	reduced incidence of VAP; slight reduction in length of stay in ICU; no difference in mortality	Kotzampassi et al. [44], 2006
L. plantarum	patients in ICU	slightly reduced incidence of VAP (not significant) and reduced incidence of *Pseudomonas* infections; slight reduction in length of stay in ICU; no difference in mortality	Klarin et al. [45], 2008

ICU = Intensive care unit; VAP = ventilatory-associated pneumonia.

The occurrence of ARI in at-risk subjects could have a recurrent and/or severe course that could significantly affect quality of daily life and, in selected cases, their survival. The role of probiotics in preventing ARI in susceptible subjects such as elderly patients, intensive care unit patients, athletes, and soldiers during strenuous training and stressed individuals has been explored in few trials. Elderly persons have frequent and severe community-acquired respiratory infections. A multicenter study in more than a 1,000 aging patients demonstrated that the daily consumption of a diary product containing the probiotic strain *L. casei* DN-114001 reduced the duration of common infectious diseases, and this effect was more evident for URTI (i.e. rhinopharyngitis) [27].

Since the prevalence of infections and subsequent mortality is high in critically ill patients, probiotics have also been tested in intensive care units. Oral administration of *L. casei rhamnosus* delayed respiratory tract colonization and infection by *P. aeruginosa* in intensive care unit patients [28]. In addition a trend toward reduction of ventilator-associated infections in intubated subjects has been shown. In a meta-analysis, the use of probiotics in this cluster of patients was associated with a lower incidence of ventilator-associated pneumonia compared to controls [29]. However, the efficacy of probiotic supplementation in reducing nosocomial infections in high-risk patients (including ventilator-associated infections) is still far from being convincing, due to the relatively low quality of available evidence and the contradictory results in recent trials [30].

Athletes and subjects under strenuous exercise, such as soldiers or marathon runners during training, seem to have no benefit on respiratory tract infections during administration of different probiotic formulations [31–33]. In a recent study, the regular ingestion of *L. casei* Shirota was beneficial in reducing the frequency of URTI in an athletic cohort, which may be related to better maintenance of saliva IgA levels during winter training and competitions [34]. A daily assumption of *L. casei* DN-114001 in stressed individuals, such as shift workers, resulted in a reduction of common infections of about 25%, and in a prolongation of the time to the first occurrence. Finally, the number of infections was reduced in the subgroup of smokers, too [35].

It should be considered that probiotics have major advantages including the low risk of adverse events, the lack of interaction with other therapies, the relatively low cost and – last but not least – the fact that they are perceived as a natural remedy rather than a polluting clinical intervention. However, the effect of probiotics on the incidence of common cold/winter diseases episodes in adults is still unclear and is only supported by a relatively smaller body of evidence compared to childhood.

The Role of Probiotics in the Prevention of Death in Preterm Infants

We will briefly discuss this highly sensitive yet potential application that has triggered a strong debate in the scientific community. Preterm infants are at increased risk of developing severe infections such as NEC, sepsis, and respiratory infections. Probiotics have been reported to reduce the all-cause mortality and specifically NEC in preterm infants by more than half [36]. Interestingly, different probiotics reduced not only the risk of NEC, but also the mortality due to other causes, including sepsis and death due to all causes, suggesting a role of probiotics in the modulation of immune system probably through an early modification of gut flora. Routine administration of probiotics to preterm infants has been proposed [37], with the strong recommendation of a combination containing a *Lactobacillus* and at least one *Bifidobacterium* species. However, a systematic review of available evidence regarding nutritional support in neonates at risk of NEC by the American Society of Parenteral and Enteral Nutrition (ASPEN) concluded that there are insufficient data to recommend the use of probiotics in infants at risk for NEC [38]. A similar conclusion was reached by European scientists [39]. The debate is still ongoing, but this data opened a burning question due to the clinical importance and consequences of what is being discussed.

Conclusions

An increasing body of evidence supports the hypothesis that modifications of intestinal microflora through the administration of probiotics may protect children and adults from respiratory tract infections through the modulation of immune response. However, the results are often inconclusive and in part inconsistent. This may be due to the heterogeneity of populations enrolled, the probiotic strains used, their dosage and duration of treatment, parameters of efficacy, and outcomes. Definitions of conditions and settings may affect the results as well and need to be carefully considered and described [9].

In children, promising results in preventing ARI were obtained with both lactobacilli and bifidobacteria, alone or in combination with other strains, and particularly in conditions of increased exposure to diseases. The use of probiotics to reduce the incidence of ARI in healthy subjects is promising and may have a great impact in terms of public health. Effective prevention of frequent although mild diseases with probiotics may significantly reduce both society and family costs. However, the use of probiotics in subjects at high risk of developing an ARI, such as CF patients, and eventually in preterm infants is highly important and opens an entirely new area of investigations into very severe conditions in which probiotics may play an important role in reducing the mortality rate and improving quality of life.

References

1 Lu N, Samuels ME, Shi L, Baker SL, Danders JM: Child day care risks of common infectious diseases revisited. Child Car Health Dev 2004;30:361–368.

2 Welliver R, McLaughing S: Unique epidemiology of nosocomial infection in children's hospital. Am J Dis Child 1984;138:131–135.

3 Polz M, Jablonski L: Nosocomial infections in children's hospital: a retrospective study. J Hyg Epidemiol Microbiol Immunol 1986;30:149–153.

4 Roos K, Hakansson EG, Holm S: Effect of recolonization with 'interfering' alfa streptococci on recurrences of acute and secretory otitis media in children: randomized placebo controlled trial. BMJ 2001;322:1–4.

5 Roos K, Holm SE, Grahn E, Lind L: Alpha-streptococcias supplementary treatment of recurrent streptococcal tonsillitis: a randomized placebo-controlled study. Scand J Infect Dis 1993;25:31–35.

6 Falck G, Grahn-Håkansson E, Holm SE, Roos K, Lagergren L: Tolerance and efficacy of interfering alpha-streptococci in recurrence of streptococcal-pharyngotonsillitis: a placebo-controlled study. Acta Otolaryngol 1999;119:944–948.

7 Glück U, Gebbers JO: Ingested probiotics reduce nasal colonization with pathogenic bacteria (*Staphylococcus aureus, Streptococcus pnumoniae*, and β-hemolytic *streptococci*). Am J Clin Nutr 2003;77:517–520.

8 Hao Q, Lu Z, Dong BR, Huang CQ, Wu T: Probiotics for preventing acute upper respiratory tract infections. Cochrane Database Syst Rev 2011;9:CD006895.

9 Koletzko B, Szajewska H, Ashwell M, Shamir R, Aggett P, Baerlocher K, Noakes P, Braegger C, Calder P, Campoy Folgoso C, Colomb V, Decsi T, Domellöf M, Dupont C, Fewtrell M, van Goudoever JB, Michaelsen KF, Mihatsch W, Guarino A, Koletzko S, Rigo J, Turck D, Taminiau J, Early Nutrition Academy and European Society for Paediatric Gastroenterology, Hepatology and Nutrition Committee on Nutrition – Consensus Group on Outcome Measures Made in Paediatric Enteral Nutrition Clinical Trials: Documentation of functional and clinical effects of infant nutrition: setting the scene for COMMENT. Ann Nutr Metab 2012;60:222–232.

10 Taipale T, Pienihäkkinen K, Isolauri E, Larsen C, Brockmann E, AlanenP, Jokela J, Söderling E: *Bifidobacterium animalis* subsp. *lactis* BB-12 in reducing the risk of infections in infancy. Br J Nutr 2011;105: 409–416.

11 Maldonado J, Cañabate F, Sempere L, Vela F, Sánchez AR, Narbona E, López-Huertas E, Geerlings A, Valero AD, Olivares M, Lara-Villoslada F: Human milk probiotic *Lactobacillus fermentum* CECT5716 reduces the incidence of gastrointestinal and upper respiratory tract infections in infants. J Pediatr Gastroenterol Nutr 2012;54:55–61.

12 Leyer GJ, Li S, Mubasher ME, Reifer C, Ouwehand AC: Probiotic effects on cold and influenza-like symptom incidence and duration in children. Pediatrics 2009;124:e172–e179.

13 Weizman Z, Asli G, Alsheikh A: Effect of a probiotic infant formula on infections in child care centers: comparison of two probiotic agents. Pediatrics 2005; 115:5–9.

14 Merenstein D, Murphy M, Fokar A, Hernandez RK, Park H, Nsouli H, Sanders ME, Davis BA, Niborski V, Tondu F, Shara NM: Use of a fermented dairy probiotic drink containing *Lactobacillus casei* (DN-114 001) to decrease the rate of illness in kids: the DRINK study. A patient-oriented, double-blind, cluster-randomized, placebo-controlled, clinical trial. Eur J Clin Nutr 2010;64:669–677.

15 Hatakka K, Savilahti E, Pönkä A, Meurman JH, Poussa T, Näse L, Saxelin M, Korpela R: Effect of long term consumption of probiotic milk on infections in children attending day care centres: double blind, randomised trial. BMJ 2001;322:1327.

16 Hojsak I, Snovak N, Abdović S, Szajewska H, Misak Z, Kolacek S: Lactobacillus GG in the prevention of gastrointestinal and respiratory tract infections in children who attend day care centers: a randomized, double-blind, placebo-controlled trial. Clin Nutr 2010;29:312–316.

17 Hojsak I, Abdović S, Szajewska H, Milosević M, Krznarić Z, Kolacek S: Lactobacillus GG in the prevention of nosocomial gastrointestinal and respiratory tract infections. Pediatrics 2010;125:e1171–e1177.

18 Niittynen L, Pitkäranta A, Korpela R: Probiotics and otitis media in children. Int J Pediatr Otorhinolaryngol 2012;76:465–470.

19 Bruzzese E, Raia V, Spagnuolo MI, Volpicelli M, De Marco G, Maiuri L, Guarino A: Effect of Lactobacillus GG supplementation on pulmonary exacerbation in patients with cystic fibrosis: a pilot study. Clin Nutr 2007;26:322–328.

20 Weiss B, Bujanover Y, Yahav Y, Vilozni D, Fireman E, Efrati O: Probiotic supplementation affects pulmonary exacerbations in patients with cystic fibrosis: a pilot study. Pediatr Pulmonol 2010;45:536–540.

21 Fendrick AM, Saint S, Brook I, Jacobs MR, Pelton S, Sethi S: Diagnosis and treatment of upper respiratory tract infections in the primary care setting. Clin Ther 2001;23:1683–1706.

22 Berggren A, Lazou Ahrén I, Larsson N, Önning G: Randomised, double-blind and placebo-controlled study using new probiotic lactobacilli for strengthening the body immune defence against viral infections. Eur J Nutr 2011;20:3–10

23 Pregliasco F Anselmi G, Fonte L, Giussani F, Schieppati S, Soletti L: A new change of preventing winter diseases by the administration of symbiotic formulations. J Clin Gastroenterol 2008;42(Suppl 3 Pt 2):S224–S233.

24 de Vrese M, Winkler P, Rautenberg, Harder T, Noah C, Laue C, Ott S, Hampe J, Schreiber S, Heller K, Schrezenmeir J: Probiotic bacteria reduced duration and severity but not the incidence of common cold episodes in a double blind, randomized controlled trial. Vaccine 2006;24:6670–6674.

25 Rizzardini G, Eskesen D, Calder PC, Capetti A, Jespersen L, Clerici M: Evaluation of the immune benefit of two probiotics strains *Bifidobacterium animalis* ssp. lactis, BB-12® and *Lactobacillus paracasei* ssp. *paracasei*, *L. casei* 431® in an influenza vaccination model: a randomized, double blind, placebo-controlled study. Br J Nutr 2012;107:876–884.

26 Davidson LE, Fiorino AM, Snydman DR, Hibberd PL: Lactobacillus GG as an immune adjuvant for live-attenuated influenza vaccine in healthy adults: a randomized double-blind placebo-controlled trial. Eur J Clin Nutr 2011;65:501–507.

27 Guillemard E, Tondu F, Lacoin F, Schrezenmeir J: Consumption of a fermented dairy product containing the probiotic *Lactobacillus casei* DN 114 001 reduces the duration of respiratory infections in the elderly in a randomized controlled trial. Br J Nutr 2010;103:58–68.

28 Forestier C, Guelon D, Cluytens V, Gillart T, Sirot J, De Champs C: Oral probiotic and prevention of *Pseudomonas aeruginosa* infections: a randomized, double-blind, placebo-controlled pilot study in intensive care unit patients. Crit Care 2008;12:R69.

29 Siempos II, Ntaidou TK, Falagas ME: Impact of the administration of probiotics on the incidence of ventilator-associated pneumonia: a meta-analysis of randomized controlled trials. Crit Care Med 2010; 38:954–962.

30 Oudhuis GJ, Bergmans D, Verbon A: Probiotcs for prevention of nosocomial infectious: efficacy and adverse effects. Curr Opin Crit Care 2011;17:487–492.

31 Kekkonen RA, Vasankari TJ, Vuorimaa T, Haahtela T, Julkunen I, Korpela R: The effect of probiotics on respiratory infections and gastrointestinal symptoms during training in marathon runners. Int J Sport Nutr Excer Metab 2007;17:352–363.

32 West NP, Pyne DB, Cripps AW, Hopkins WG, Eskesen DC, Jairath A, Christophersen CT, Conlon MA, Fricker PA: *Lactobacillus fermentum* (PCC) supplementation and gastrointestinal and respiratory-tract illness symptoms: a randomized control trial in athletes. Nutr J 2011;10:30.

33 Tiollier E, Chennaoui M, Gomez-Merino D, et al: Effect of a probiotics supplementation on respiratory infections and immune and hormonal parameters during intense military training. Mil Med 2007;172: 1006–1011.

34 Gleeson M, Bishop NC, Oliveira M, Tauler P: Daily probiotic's (*Lactobacillus casei* Shirota) reduction of infection incidence in athletes. Int J Sport Nutr Exerc Metab 2011;21:55–64.

35 Guillemard E, Tanguy J, Flavigny A, de la Motte S, Schrezenmeir J: Effects of consumption of a fermented dairy product containing the probiotic *Lactobacillus casei* DN-114 001 on common respiratory and gastrointestinal infections in shift workers in a randomized controlled trial. J Am Coll Nutr 2010;29: 455–468.

36 Deshpande G, Rao S, Patole S, Bulsara M: Updated meta-analysis of probiotics for preventing necrotising enterocolitis in preterm neonates. Pediatrics 2010;125:921–930.

37 Deshpande CG, Rao SC, Keil AD, Patole SK: Evidence-based guidelines for use of probiotics in preterm neonates. BMC Med 2011;9:92–105.

38 Fallon EM, Nehra D, Potemkin AK, Gura KM, Simpser E, Compher C, American Society for Parenteral and Enteral Nutrition (A.S.P.E.N.) Board of Directors, Puder M: A.S.P.E.N. clinical guidelines: nutrition support of neonatal patients at risk for necrotizing enterocolitis. JPEN J Parenter Enteral Nutr 2012;36:506–523.

39 Mihatsch WA, Braegger CP, Decsi T, Kolacek S, Lanzinger H, Mayer B, Moreno LA, Pohlandt F, Puntis J, Shamir R, Stadtmüller U, Szajewska H, Turck D, van Goudoever JB: Critical systematic review of the level of evidence for routine use of probiotics for reduction of mortality and prevention of necrotizing enterocolitis and sepsis in preterm infants. Clin Nutr 2012;31:6–15.

40 Glück U, Gebbers JO: Ingested probiotics reduce nasal colonization with pathogenic bacteria (Staphylococcus aureus, Streptococcus pneumoniae, and beta-hemolytic streptococci). Am J Clin Nutr 2003;77:517–520.

41 Makino S, Ikegami S, Kume A, Horiuchi H, Sasaki H, Orii N: Reducing the risk of infection in the elderly by dietary intake of yoghurt fermented with Lactobacillus delbrueckii ssp. bulgaricus OLL1073R-1. Br J Nutr 2010;104:998–1006.

42 Knight DJ, Gardiner D, Banks A, Snape SE, Weston VC, Bengmark S, Girling KJ: Effect of synbiotic therapy on the incidence of ventilator associated pneumonia in critically ill patients: a randomised, double-blind, placebo-controlled trial. Intensive Care Med 2009;35:854–861.

43 Spindler-Vesel A, Bengmark S, Vovk I, Cerovic O, Kompan L: Synbiotics, prebiotics, glutamine, or peptide in early enteral nutrition: a randomized study in trauma patients. J Parenter Enteral Nutr 2007;31:119–126.

44 Kotzampassi K, Giamarellos-Bourboulis EJ, Voudouris A, Kazamias P, Eleftheriadis E: Benefits of a synbiotic formula (Synbiotic 2000Forte) in critically Ill trauma patients: early results of a randomized controlled trial. World J Surg 2006;30:1848–1855.

45 Klarin B, Molin G, Jeppsson B, Larsson A: Use of the probiotic Lactobacillus plantarum 299 to reduce pathogenic bacteria in the oropharynx of intubated patients: a randomised controlled open pilot study. Critical Care 2008;12:R136.

Eugenia Bruzzese, MD
Section of Pediatrics, Department of Translational Medical Science
University of Naples 'Federico II'
Via Sergio Pansini 5, 80131 Naples (Italy)
E-Mail eugbruzz@unina.it

Guarino A, Quigley EMM, Walker WA (eds): Probiotic Bacteria and Their Effect on Human Health and Well-Being.
World Rev Nutr Diet. Basel, Karger, 2013, vol 107, pp 151–160 (DOI: 10.1159/000345751)

Are Probiotic Effects Dose-Related?

Tiffany J. Patton · Stefano Guandalini

Department of Pediatrics, Section of Gastroenterology, University of Chicago Comer Children's Hospital,
Chicago, Ill., USA

Abstract

Probiotics are increasingly used in order to help prevent or treat a growing number of conditions, especially gastrointestinal tract disorders. Indeed, a large number of clinical trials, many properly randomized and controlled, have shown significant efficacy in a number of gastrointestinal conditions. However, studies have been very heterogeneous and conclusions are often hard to draw. An issue which has been grossly neglected appears to be that of pharmacokinetics. In fact, studies addressing any type of dose-effect relationship are extremely limited, and drawing conclusions on the relationship between dose and effect for any probiotic species in the context of any specific disorder is close to impossible. This chapter analyzes the available evidence on this aspect, looking at areas such as acute infectious diarrhea, prevention of antibiotic-associated diarrhea, necrotizing enterocolitis, and irritable bowel syndrome. As we acknowledge the paucity of data available on the pharmacokinetics of probiotics and aim to understand more about their mechanisms of action, caution must also be exerted on possible unsafe levels. It is evident that much more solid evidence needs to be provided on the dose-effect relationship of probiotics by analyzing single preparations in rigorously controlled studies that must also aim at establishing possible unsafe or risky levels.

A recent survey among physicians in the UK [1] showed that 70% of respondents recommended or prescribed probiotics to their patients, including 53% of surgeons and 80% of gastroenterologists. The most popular indications were irritable bowel syndrome (IBS; 70% of prescribers) and pouchitis (67% of prescribers). Yet, this enthusiasm for probiotics, which is likely to be worldwide and in large part based on the vast amount of data that has accumulated over the past decade or two, does not appear to be based on a sound understanding of their pharmacokinetics, which remains poorly understood.

In 2001, an expert group of the WHO/FAO defined probiotics as 'live microorganisms that when administered in adequate amounts confer a health benefit on the host' [2]. Thus, it is already apparent that in such a definition, which is currently widely accepted by experts in the field, the existence of a dose-effect relationship ('in adequate amounts') is implied as relevant to the health benefit. In spite of this, very scanty data are available that investigate such a relationship, notwithstanding an enormous increase in the number of significant clinical trials throughout the world highlighting the utility and efficacy of probiotics in a wide variety of medical conditions. Among these conditions are gastrointestinal diseases (acute diarrhea, IBS, inflammatory bowel disease, *Helicobacter pylori* gastritis, and necrotizing enterocolitis) as well as extraintestinal conditions such as nephrolithiasis, urinary tract infections, vaginosis, and atopic illnesses. In fact, while many of these studies have assessed probiotic efficacy compared to placebo or other alternative therapies, dose comparisons of the effects of the administration of living microbes in the ingested formulation – defined as colony-forming units (CFU) either per dose or per ml – have been infrequent.

A variety of factors need to be considered when discussing the dose efficacy of probiotics. To begin with, it is generally thought that probiotics, to be effective, have to be ingested in numbers sufficient to survive the normal gastrointestinal milieu of gastric acid, bile salts, and pancreatic enzymes, and reach 'adequate' concentrations in the intestinal lumen, where they are supposed to remain alive and possibly colonize. However, it has never been satisfactorily shown that any probiotic is able to multiply or colonize in the small intestine; furthermore, to what extent they may do so in the colon is still unclear and it may well be strain dependent. Thus, assuming that probiotics have a limited capacity of colonization in the intestine and need to remain viable, the choice of a proper probiotic concentration in the formulation to be ingested must be based on a combination of factors. Most of these factors are poorly understood, but include the unavoidable dilution they will undergo while transiting through the gastrointestinal tract, the capacity of the strain to survive to gastric and pancreatic secretion, and the expected health benefit of that specific strain (or strains) in that specific condition and for that specific host.

Table 1 summarizes some of the published data concerning the viability of various probiotics in humans. It can be seen that different strains were tested in a wide range of concentrations (from 10^7 to 10^{11} CFU/g), and the parameters checked (ileal, colonic, fecal concentrations, and fecal recovery) show a likewise varying degree of viability, with recoveries varying, for example, between 1 and 92%.

Several studies have assessed the efficacy of various probiotics in gastrointestinal disease, with the most well-known probiotics being lactic acid bacteria and the yeast *Saccharomyces boulardii*. It is beyond the scope of this chapter to review them in detail. Therefore, we will focus instead on areas where some inferences about any dose-effect relationship can be drawn, even though in most cases such conclusions are based on comparisons between different studies rather than a comparison of various doses of the same preparation in a single study.

Acute Infectious Diarrhea

In a Web-based survey, Weizman [3] found in 2010 that many pediatric gastroenterologists worldwide, in spite of considering the use of probiotics effective in childhood acute diarrhea, did not use them consistently, owing to lack of appropriate

Table 1. Viability of various probiotic strains in humans

Probiotic strains	Dose administered, CFU/g	Ileal concentration, CFU/ml	Fecal concentration, CFU/g	Recovery, %	Reference
L. plantarum NCIMB 8826	10^{10}	10^8 (7%)	10^8	25–30	44
L. plantarum 299v	10^8	1–10% of dose	cecal adhesion 20–40%	25–30	45
L. fermentum KLD	10^7	10^4 (0.5%)	very low		44
L. rhamnosus GG	5×10^9	–	10^6	1–20	46
	6×10^{10}		10^7		47
L. acidophilus	fermented product	1–10% of dose	10^6	2–5	46
L. johnsonii	10^8	1–10%	10^4		46
L. johnsonii La1	10^9		10^5		48
L. johnsonii La1	10^7		10^4	30%	49
	10^9		10^5	60%	
L. casei	10^7	10% of dose	10^5		46
L. casei DN114001	10^{11}	10^8 (3.6%)	10^7		50
L. paracasei paracasei CRL 431	10^{10}/day		no detection	0	51, 52
	10^{11}		no detection	0	
	10^{12}		no detection	0	
L. paracasei A	fermented product	N/A	10^7	92%	53
L. reuteri DSM17938	8×10^8		10^5		47
	5×10^9		10^6		
	1×10^{10}		10^7		
L. salivarius 433118	fermented product	10^7	10^7		46
Lactococcus lactis MG 1363	10^7	10^5 (1.0%)	very low	0.1–2.0	44
B. stearothermophilus	10^8	10^6	10^5		44
	10^8	10^8			54
Yogurt bacteria	10^7/ml	10^5			55
L. delbrueckii bulgaricus	10^9	10^5–10^6			46
S. boulardii	1 g/day		105	30–60	56
S. boulardii	3×10^7/day		1.4×10^7		57
S. boulardii	1–3 g		8.6×10^8	0.2	58
Bifidobacteria spp.	10^{11} 4×10^9 in milk products	10^7	10^8	25–30	44, 54
B. animalis lactis BB12	10^{10}/day		10^7		51, 52
	10^{11}/day		10^7		
B. longum BB536	10^7	no detection	no detection	0	49
	10^9	no detection	no detection		

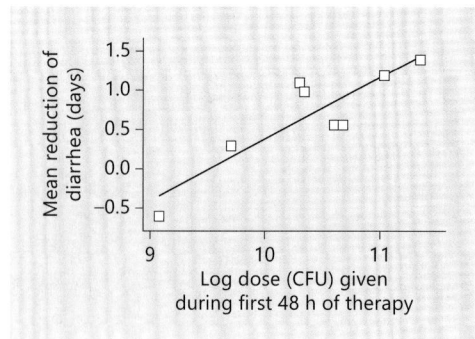

Fig. 1. Dose-effect relationship between *Lactobacillus* dose and reduction in diarrhea duration.

guidelines on the choice of probiotics and indications on doses and regimens of administration. As a result, when probiotics were utilized, dosing and duration were extremely variable.

Yet, and in spite of substantial methodological variations between the investigations, a recent review [4] analyzing the large body of evidence available on the use of probiotics in acute-onset diarrhea in children concluded that probiotics in acute infectious diarrhea were beneficial, albeit only moderately (about 1 day reduction of its duration), and depended on both the strain of microorganism and dose, with a greater effect seen with doses greater than 10^{10} to 10^{11} CFU/day. These beneficial effects were significant in watery diarrhea and viral gastroenteritis, but absent in invasive bacterial diarrhea, and were greater when probiotics were administered early in the illness [4, 5]. The use of probiotics for acute infectious diarrhea in children, and especially those found to be most effective such as *Lactobacillus* GG and *S. boulardii*, is now an accepted therapy in Europe [6, 7]. In the USA, the American Academy of Pediatrics likewise noticed in 2010 that *Lactobacillus* GG is the most effective probiotic reported to date, with an effect that is dose dependent for doses greater than 10^{10} CFU, and concluded that there is evidence to support the use of probiotics, specifically *Lactobacillus* GG, early in the course of acute infectious diarrhea to reduce the duration by 1 day [8]. Indeed, in 2002 Van Niel et al. [9] were able to show in a meta-analysis of 9 published randomized controlled studies on the efficacy of lactobacilli in acute diarrhea in children that a dose-response effect was present, with daily doses equal to or above 10^{10} CFU (fig. 1) needed to achieve significant reduction of diarrhea duration.

Prevention of Antibiotic-Associated Diarrhea

One of the major complications of treatment with antimicrobial agents is antibiotic-associated diarrhea (AAD), occurring in 5–25% of patients [10]. In addition, infection with *Clostridium difficile* is a well-recognized major cause of AAD, believed to be re-

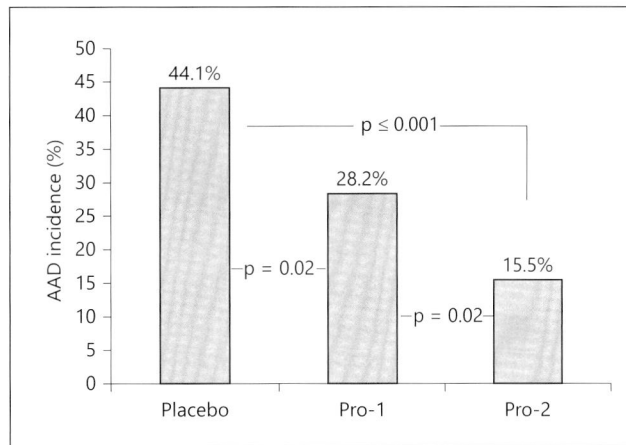

Fig. 2. Effects of two different doses of a probiotic preparation (*L. acidophilus* CL1285 and *L. casei* LBC80R) on the incidence of antibiotic-associated diarrhea. Pro-1 (n = 85) patients received 1 capsule per day, while Pro-2 patient (n = 86) received 2 capsules per day. Each probiotic capsule contained 50 billion CFU of live organisms.

sponsible for 15–25% of AAD cases [11]. Already in the mid-90's, Saccharomyces boulardii and Lactobacillus GG had been found effective in controlling, respectively, *C. difficile*-enteropathy and colitis [15, 16]. One of the very few studies directly comparing different doses of the same probiotic preparation in the same setting was published in 2010 [12]. The study compared two doses of a proprietary formulation of probiotics (*Lactobacillus acidophilus* CL1285 and *Lactobacillus casei* LBC80R) on the prophylaxis of AAD and *C. difficile*-associated diarrhea in adult patients. In this single-center, randomized, double-blind, placebo-controlled dose-ranging study, the authors randomized 255 adult inpatients to one of three groups: two probiotic capsules per day (Pro-2, n = 86), one probiotic capsule and one placebo capsule per day (Pro-1, n = 85), or two placebo capsules per day (n = 84). Each probiotic capsule contained 50 billion CFU of live organisms. While each probiotic group had a lower AAD incidence versus placebo (44.1%), Pro-2 patients had a lower AAD incidence versus Pro-1 (15.5 vs. 28.2%; fig. 2).

In a very recent large meta-analysis, Hempel et al. [13] included 82 randomized controlled trials, the majority of which used *Lactobacillus*-based interventions alone or in combination with other genera; strains were poorly documented. The pooled relative risk in an analysis that included almost 12,000 participants indicated a statistically significant association of probiotic administration with reduction in AAD [relative risk: 0.58 (95% CI: 0.50–0.68), $p < 0.001$; I(2): 54%; risk difference: –0.07 (95% CI: –0.10 to –0.05); number needed to treat: 13 (95% CI: 10.3–19.1)] in trials reporting on the number of patients with AAD. However, the authors noticed a significant heterogeneity in pooled results and commented that the evidence is insufficient to determine whether this association varies systematically by population, antibiotic characteristic, or probiotic preparation.

Previously, Johnston et al. [14] assessed the dose effect of probiotics for the prevention of pediatric AAD. They reviewed 16 studies with over 3,400 participants (ages 2 weeks to 17 years). The trials included simultaneous antibiotic and probiotic administration with *Bacillus* spp., *Bifidobacterium* spp., *Lactobacillus* spp., *Lactococcus* spp., *Leuconostoc cremoris*, *Saccharomyces* spp., or *Streptococcus* spp., alone or in combination. Among the investigated strains, again *Lactobacillus* GG and *S. boulardii* appeared the most effective. In a subgroup analysis, they analyzed low-dose (<5 billion CFU/day) versus high-dose (≥5 billion CFU/day) effects on AAD. In studies providing ≥5 billion CFU/day, the pooled incidence of AAD in the probiotic group was 8% compared to 22% in the active, placebo, or no treatment group (RR: 0.40; 95% CI: 0.29–0.55). Additionally, in studies providing less than 5 billion CFU/day, the pooled incidence of AAD in the probiotic group was 8% compared to 11% in the active, placebo, or no treatment control group (RR: 0.80; 95% CI: 0.53–1.21).

Necrotizing Enterocolitis

An extremely interesting potential application for probiotics is for necrotizing enterocolitis, an often lethal disease affecting premature infants. Several clinical trials with various probiotic preparations have been done with exciting results, but it is impossible to identify optimal strain(s) from them, and especially any firm conclusions about their dosing. In spite of this, evidence-based guidelines for their use in preterm neonates have been published [17] indicating a daily dose of $1.5–3.0 \times 10^9$ CFU/day for extremely-low-birth-weight preterms as appropriate, with the additional recommendation of starting the supplementation along with enteral feedings as early as possible. When comparing such doses for extremely-low-birth-weight preterms to doses utilized in older children for other gastrointestinal conditions (e.g. gastroenteritis), they appear to be very high.

A meta-analysis on this topic – not addressing the issue of dose-effect – similarly concluded that enteral supplementation with probiotics does indeed prevent severe necrotizing enterocolitis as well as mortality from all causes in preterm infants [18]. The study even concluded that the evidence supports changing clinical practice. Of interest, however, specific safety data on extremely-low-birth-weight preterms were not offered. In this regard, it is worth noting that a recent retrospective report [19] in preterm babies comparing a historic cohort with a more recent one in which the authors routinely used a probiotic (*Lactobacillus rhamnosus* Lcr35) at the relatively low dose of 2×10^8 CFU/12 h showed a reduced rate of necrotizing enterocolitis (OR: 0.20), reduced mortality (OR: 0.46), and reduced incidence of late-onset sepsis (OR: 0.60). The cohort on probiotics also achieved full enteral feedings significantly earlier. Thus, for some strains it appears that doses well below those used so far might be adequate and are likely to offer a wider margin of safety, something clearly desirable since concerns about the use of probiotics in such premature babies have been raised [20].

Irritable Bowel Syndrome

IBS is a highly prevalent functional gastrointestinal disorder. Although the etiology of IBS remains elusive, there is growing recognition of the role played by intestinal infections and disturbances of the colonic microflora in the genesis of this condition [21–23]. Furthermore, it has been shown that modified commensal gut flora may lead to mucosal inflammation. Several changes such as an increase in mucosal cellularity, modified proinflammatory/anti-inflammatory cytokine balance, and disordered neurotransmission have been observed in postinfectious IBS, thought to account for about 30% of all IBS cases. Hence, the use of probiotics has been proposed with evidence of effectiveness both in adults and in pediatric patients.

Clarke et al. [24] recently analyzed the evidence to support the use of probiotics – lactic acid bacteria – in this disorder. Of the 42 trials evaluated which examined the efficacy of lactic acid bacteria in IBS, 34 reported beneficial effects in at least one of the endpoints or symptoms examined, albeit with tremendous variation in both the magnitude of effect and the choice of outcome under consideration. However, the authors also noticed again one of the usual pitfalls of studies with probiotics: deficits of trial design and execution relating to strain selection, optimum dosage, mode of action, safety, and long-term tolerability. In spite of these negative remarks, the rather solid conceptual basis for the use of probiotics in IBS [25] and the generally positive outcome of the studies performed allowed the authors to conclude that the 'future holds much promise for the use of lactic acid bacteria in the treatment of irritable bowel syndrome.'

A previous meta-analysis including all strains of probiotics [26] included 1,650 patients from 19 randomized controlled trials. Even in the presence of significant heterogeneity, the analysis showed that probiotics were statistically significantly better than placebo (RR of IBS not improving = 0.71; 95% CI: 0.57–0.88) with a number needed to treat of 4 (95% CI: 3–12.5). The authors concluded that probiotics appear to be efficacious in IBS, but the magnitude of the benefit and the most effective species and strain are uncertain.

Of note, even though any conclusions on the dose-effect response cannot be drawn on this topic due to the lack of studies addressing this aspect, there are suggestions that a mixture of probiotics (hence, a much higher final CFU/day) may be more effective than single strains in IBS [27]. Indeed, our recent investigation [28] with VSL#3 (a mixture of 8 different strains) in children with IBS showed this highly concentrated preparation to be safe and effective.

Is Too Much of a Good Thing a Bad Thing?

One should also be aware of the 'other side' of the dose-effect relationship: 'Is there a dose that is too high, thus causing adverse effects?' In reality, probiotics have been administered safely to individuals with immunoinflammatory disorders such as atopy

[29] and Crohn's disease [30] as well as those with HIV and immunosuppression [31, 32]. However, they certainly may exert a stimulatory effect on the intestinal immune system that can result in proinflammatory stimuli (e.g. fecal calprotectin, a marker of inflammation, is increased after the administration of *Lactobacillus reuteri* DSM 17938 [33]) and it is not known if there is a sort of a 'therapeutic window' for specific probiotic strains above which the threshold for safety may be crossed.

Besides the obvious evidence provided by the rare case reports of bacteremia [34] [35, 36] or fungemia [37–39] occurring after 'regular' doses of probiotic supplementation in vulnerable individuals, some reports such as the increased risk of mortality found in patients with severe pancreatitis receiving probiotics [40] have alerted the scientific community to this possibility. Recent animal studies have also shown the risk of excessive doses. In fact, administration of high-dose *L. rhamnosus* (10^{12} CFU/day) resulted in an increased incidence of diarrhea in piglets before challenge with an enterotoxigenic *Escherichia coli* strain (F4 + ETEC) [41]. Similarly, a recent trial of piglets challenged with F4 + ETEC showed that dietary addition of *Lactobacillus rhamnosus* GG (10^{10} CFU/day) failed to prevent F4 + ETEC infection, possibly due to the greater degree of stress caused by the challenge [42]. In fact, it has been reported that high-dose (10^{10} CFU/l) but not low-dose *Lactobacillus rhamnosus* GG has proinflammatory properties in Caco-2 cells by markedly increasing IL-8 production [43].

Therefore, it is evident that much more solid evidence needs to be provided on the pharmacokinetics of probiotics, analyzing single preparations in rigorously controlled studies that must also focus on the identification of possible unsafe or risky levels.

References

1 Cordina C, Shaikh I, Shrestha S, Camilleri-Brennan J: Probiotics in the management of gastrointestinal disease: analysis of the attitudes and prescribing practices of gastroenterologists and surgeons. J Dig Dis 2011;12:489–496.

2 Food and Agriculture Organization and World Health Organization Expert Consultation. Evaluation of Health and Nutritional Properties of Powder Milk and Live Lactic Acid Bacteria. Córdoba, WHO-FAO, 2001.

3 Weizman Z: Probiotics use in childhood acute diarrhea: a Web-based survey. J Clin Gastroenterol 2011; 45:426–428.

4 Wolvers D, Antoine JM, Myllyluoma E, Schrezenmeir J, Szajewska H, Rijkers GT: Guidance for substantiating the evidence for beneficial effects of probiotics: prevention and management of infections by probiotics. J Nutr 2010;140:698S–712S.

5 Allen SJ, Martinez EG, Gregorio GV, Dans LF: Probiotics for treating acute infectious diarrhoea. Cochrane Database Syst Rev 2010;11:CD003048.

6 Guarino A, Albano F, Ashkenazi S, et al: European Society for Paediatric Gastroenterology, Hepatology, and Nutrition/European Society for Paediatric Infectious Diseases evidence-based guidelines for the management of acute gastroenteritis in children in Europe. J Pediatr Gastroenterol Nutr 2008;46(Suppl 2):S81–S122.

7 Guandalini S: Acute diarrhea in children in Europe: do we know how to treat it? J Pediatr Gastroenterol Nutr 2008;46(Suppl 2):S77–S80.

8 Thomas DW, Greer FR: Probiotics and prebiotics in pediatrics. Pediatrics 2010;126:1217–1231.

9 Van Niel CW, Feudtner C, Garrison MM, Christakis DA: Lactobacillus therapy for acute infectious diarrhea in children: a meta-analysis. Pediatrics 2002; 109:678–684.

10 Bergogne-Berezin E: Treatment and prevention of antibiotic associated diarrhea. Int J Antimicrob Agents 2000;16:521–526.

11 Barbut F, Petit JC: Epidemiology of *Clostridium difficile*-associated infections. Clin Microbiol Infect 2001;7:405–410.

12 Gao XW, Mubasher M, Fang CY, Reifer C, Miller LE: Dose-response efficacy of a proprietary probiotic formula of *Lactobacillus acidophilus* CL1285 and *Lactobacillus casei* LBC80R for antibiotic-associated diarrhea and *Clostridium difficile*-associated diarrhea prophylaxis in adult patients. Am J Gastroenterol 2010;105:1636–1641.

13 Hempel S, Newberry SJ, Maher AR, et al: Probiotics for the prevention and treatment of antibiotic-associated diarrhea: a systematic review and meta-analysis. JAMA 2012;307:1959–1969.

14 Johnston BC, Goldenberg JZ, Vandvik PO, Sun X, Guyatt GH: Probiotics for the prevention of pediatric antibiotic-associated diarrhea. Cochrane Database Syst Rev 2011;11:CD004827.

15 Buts JP, Corthier G, Delmee M: *Saccharomyces boulardii* for *Clostridium difficile*-associated enteropathies in infants. J Pediatr Gastroenterol Nutr 1993;16:419–425.

16 Biller JA, Katz AJ, Flores AF, Buie TM, Gorbach SL: Treatment of recurrent *Clostridium difficile* colitis with Lactobacillus GG. J Pediatr Gastroenterol Nutr 1995;21:224–226.

17 Deshpande GC, Rao SC, Keil AD, Patole SK: Evidence-based guidelines for use of probiotics in preterm neonates. BMC Med 2011;9:92.

18 Alfaleh K, Anabrees J, Bassler D, Al-Kharfi T: Probiotics for prevention of necrotizing enterocolitis in preterm infants. Cochrane Database Syst Rev 2011; 3:CD005496.

19 Bonsante F, Iacobelli S, Gouyon JB: Routine probiotic use in very preterm infants: retrospective comparison of two cohorts. Am J Perinatol 2012, E-pub ahead of print.

20 Beattie LM, Hansen R, Barclay AR: Probiotics for preterm infants: confounding features warrant caution. Pediatrics 2010;126:e742–e743, author reply e743–e745.

21 Cuomo R, Savarese MF, Gargano R: Almost all irritable bowel syndromes are post-infectious and respond to probiotics: consensus issues. Dig Dis 2007; 25:241–244.

22 Wilhelm SM, Brubaker CM, Varcak EA, Kale-Pradhan PB: Effectiveness of probiotics in the treatment of irritable bowel syndrome. Pharmacotherapy 2008; 28:496–505.

23 Kajander K, Myllyluoma E, Rajilic-Stojanovic M, et al: Clinical trial: multispecies probiotic supplementation alleviates the symptoms of irritable bowel syndrome and stabilizes intestinal microbiota. Aliment Pharmacol Ther 2008;27:48–57.

24 Clarke G, Cryan JF, Dinan TG, Quigley EM: Review article: probiotics for the treatment of irritable bowel syndrome – focus on lactic acid bacteria. Aliment Pharmacol Ther 2012;35:403–413.

25 Dahlqvist G, Piessevaux H: Irritable bowel syndrome: the role of the intestinal microbiota, pathogenesis and therapeutic targets. Acta Gastroenterol Belg 2011;74:375–380.

26 Moayyedi P, Ford AC, Talley NJ, et al: The efficacy of probiotics in the treatment of irritable bowel syndrome: a systematic review. Gut 2010;59:325–332.

27 Chapman CM, Gibson GR, Rowland I: Health benefits of probiotics: are mixtures more effective than single strains? Eur J Nutr 2011;50:1–17.

28 Guandalini S, Magazzu G, Chiaro A, et al: VSL#3 improves symptoms in children with irritable bowel syndrome: a multicenter, randomized, placebo-controlled, double-blind, crossover study. J Pediatr Gastroenterol Nutr 2010;51:24–30.

29 Kirjavainen PV, Salminen SJ, Isolauri E: Probiotic bacteria in the management of atopic disease: underscoring the importance of viability. J Pediatr Gastroenterol Nutr 2003;36:223–227.

30 Malin M, Suomalainen H, Saxelin M, Isolauri E: Promotion of IgA immune response in patients with Crohn's disease by oral bacteriotherapy with Lactobacillus GG. Ann Nutr Metab 1996;40:137–145.

31 Hemsworth JC, Hekmat S, Reid G: Micronutrient supplemented probiotic yogurt for HIV-infected adults taking HAART in London, Canada. Gut Microbes 2012;3:414–419.

32 Wolf BW, Wheeler KB, Ataya DG, Garleb KA: Safety and tolerance of *Lactobacillus reuteri* supplementation to a population infected with the human immunodeficiency virus. Food Chem Toxicol 1998;36: 1085–1094.

33 Mangalat N, Liu Y, Fatheree NY, et al: Safety and tolerability of *Lactobacillus reuteri* DSM 17938 and effects on biomarkers in healthy adults: results from a randomized masked trial. PLoS One 2012;7:e43910.

34 Ledoux D, Labombardi VJ, Karter D: *Lactobacillus acidophilus* bacteraemia after use of a probiotic in a patient with AIDS and Hodgkin's disease. Int J STD AIDS 2006;17:280–282.

35 Kunz AN, Noel JM, Fairchok MP: Two cases of *Lactobacillus bacteremia* during probiotic treatment of short gut syndrome. J Pediatr Gastroenterol Nutr 2004;38:457–458.

36 Abgrall S, Joly V, Derkinderen P, Decre D, Carbon C, Yeni P: *Lactobacillus casei* infection in an AIDS patient. Eur J Clin Microbiol Infect Dis 1997;16:180–182.

37 Thygesen JB, Glerup H, Tarp B: *Saccharomyces boulardii* fungemia caused by treatment with a probioticum. BMJ Case Rep 2012;2012: pii: bcr0620114412.

38 Lestin F, Pertschy A, Rimek D: Fungemia after oral treatment with *Saccharomyces boulardii* in a patient with multiple comorbidities (in German). Dtsch Med Wochenschr 2003;128:2531–2533.

39 Cherifi S, Robberecht J, Miendje Y: *Saccharomyces cerevisiae* fungemia in an elderly patient with *Clostridium difficile* colitis. Acta Clin Belg 2004;59:223–224.

40 Besselink MG, van Santvoort HC, Buskens E, et al: Probiotic prophylaxis in predicted severe acute pancreatitis: a randomised, double-blind, placebo-controlled trial. Lancet 2008;371:651–659.

41 Li XQ, Zhu YH, Zhang HF, et al: Risks associated with high-dose *Lactobacillus rhamnosus* in an *Escherichia coli* model of piglet diarrhoea: intestinal microbiota and immune imbalances. PLoS One 2012; 7:e40666.

42 Trevisi P, Casini L, Coloretti F, Mazzoni M, Merialdi G, Bosi P: Dietary addition of *Lactobacillus rhamnosus* GG impairs the health of *Escherichia coli* F4-challenged piglets. Animal 2011;5:1354–1360.

43 Zhang L, Li N, Caicedo R, Neu J: Alive and dead *Lactobacillus rhamnosus* GG decrease tumor necrosis factor-alpha-induced interleukin-8 production in Caco-2 cells. J Nutr 2005;135:1752–1756.

44 Vesa T, Pochart P, Marteau P: Pharmacokinetics of *Lactobacillus plantarum* NCIMB 8826, *Lactobacillus fermentum* KLD, and *Lactococcus lactis* MG 1363 in the human gastrointestinal tract. Aliment Pharmacol Ther 2000;14:823–828.

45 Johansson ML, Molin G, Jeppsson B, Nobaek S, Ahrne S, Bengmark S: Administration of different Lactobacillus strains in fermented oatmeal soup: in vivo colonization of human intestinal mucosa and effect on the indigenous flora. Appl Environ Microbiol 1993;59:15–20.

46 Marteau P, Shanahan F: Basic aspects and pharmacology of probiotics: an overview of pharmacokinetics, mechanisms of action and side-effects. Best Pract Res Clin Gastroenterol 2003;17:725–740.

47 Dommels YE, Kemperman RA, Zebregs YE, et al: Survival of *Lactobacillus reuteri* DSM 17938 and *Lactobacillus rhamnosus* GG in the human gastrointestinal tract with daily consumption of a low-fat probiotic spread. Appl Environ Microbiol 2009;75: 6198–6204.

48 Donnet-Hughes A, Rochat F, Serrant P, Aeschlimann JM, Schiffrin EJ: Modulation of nonspecific mechanisms of defense by lactic acid bacteria: effective dose. J Dairy Sci 1999;82:863–869.

49 Gianotti L, Morelli L, Galbiati F, et al: A randomized double-blind trial on perioperative administration of probiotics in colorectal cancer patients. World J Gastroenterol 2010;16:167–175.

50 Oozeer R, Leplingard A, Mater DD, et al: Survival of *Lactobacillus casei* in the human digestive tract after consumption of fermented milk. Appl Environ Microbiol 2006;72:5615–5617.

51 Larsen CN, Nielsen S, Kaestel P, et al: Dose-response study of probiotic bacteria *Bifidobacterium animalis* subsp *lactis* BB-12 and *Lactobacillus paracasei* subsp *paracasei* CRL-341 in healthy young adults. Eur J Clin Nutr 2006;60:1284–1293.

52 Christensen HR, Larsen CN, Kaestel P, et al: Immunomodulating potential of supplementation with probiotics: a dose-response study in healthy young adults. FEMS Immunol Med Microbiol 2006;47: 380–390.

53 Marzotto M, Maffeis C, Paternoster T, et al: *Lactobacillus paracasei* A survives gastrointestinal passage and affects the fecal microbiota of healthy infants. Res Microbiol 2006;157:857–866.

54 Pochart P, Marteau P, Bouhnik Y, Goderel I, Bourlioux P, Rambaud JC: Survival of bifidobacteria ingested via fermented milk during their passage through the human small intestine: an in vivo study using intestinal perfusion. Am J Clin Nutr 1992;55: 78–80.

55 Pochart P, Dewit O, Desjeux JF, Bourlioux P: Viable starter culture, beta-galactosidase activity, and lactose in duodenum after yogurt ingestion in lactase-deficient humans. Am J Clin Nutr 1989;49:828–831.

56 Blehaut H, Massot J, Elmer GW, Levy RH: Disposition kinetics of *Saccharomyces boulardii* in man and rat. Biopharm Drug Dispos 1989;10:353–364.

57 Elmer GW, Corthier G: Modulation of *Clostridium difficile* induced mortality as a function of the dose and the viability of the *Saccharomyces boulardii* used as a preventative agent in gnotobiotic mice. Can J Microbiol 1991;37:315–317.

58 Klein SM, Elmer GW, McFarland LV, Surawicz CM, Levy RH: Recovery and elimination of the biotherapeutic agent, *Saccharomyces boulardii*, in healthy human volunteers. Pharm Res 1993;10:1615–1619.

Stefano Guandalini, MD
Professor and Chief, Section of Pediatric Gastroenterology, Founder and Medical Director
Celiac Disease Center, University of Chicago, Comer Children's Hospital
5839 S. Maryland Ave., MC 4065, Chicago, IL 60637
E-mail sguandalini@peds.bsd.uchicago.edu

Guarino A, Quigley EMM, Walker WA (eds): Probiotic Bacteria and Their Effect on Human Health and Well-Being.
World Rev Nutr Diet. Basel, Karger, 2013, vol 107, pp 161–170 (DOI: 10.1159/000345744)

Safety of Probiotics

Iva Hojsak[a, b] · Raanan Shamir[c, d]

[a]Referral Center for Pediatric Gastroenterology and Nutrition, Children's Hospital Zagreb, and [b]University of Zagreb School of Medicine, Zagreb, Croatia; [c]Institute for Gastroenterology, Nutrition and Liver Diseases, Schneider Children's Medical Center of Israel, Petach Tikva, and [d]Sackler Faculty of Medicine, Tel Aviv University, Tel Aviv, Israel

Abstract

Probiotics are defined as live microorganisms which, when provided in sufficient amounts, have a beneficial effect on human health. In the last two decades, worldwide interest in the use of functional foods containing probiotic bacteria for disease prevention and health promotion has increased significantly, revealing new possible indications for the use of specific probiotic strains. On the other hand, increased probiotic use necessitates the need for accurate surveillance, not only for their effect, but also for their safety. Taking into account that probiotics are live microorganisms that can influence the human immune system, several safety issues need to be addressed. These include the potential for transmigration causing bacteremia and sepsis, the potential for antibiotic resistance, and other adverse effects such as those that might influence gastrointestinal physiology or immunological function. The aim of this review is to summarize available data on the safety of probiotics for human use.

Copyright © 2013 S. Karger AG, Basel

Probiotics are defined as live microorganisms which, when provided in sufficient amounts, have a beneficial effect on human health [1]. Microbes referred to as probiotics are bacteria and fungi, most commonly including species of *Lactobacillus* and *Bifidobacterium* and species of the yeast genus *Saccharomyces*. Other bacterial genera which feature probiotic strains include *Streptococcus*, *Enterococcus*, and *Bacillus*. In the last two decades, worldwide interest in the use of functional foods containing probiotic bacteria for health promotion and disease prevention has increased significantly, revealing new possible indications for the use of specific probiotic strains. In addition to the need to demonstrate evidence of the benefit, the use of each specific probiotic strain must follow a demonstrated safety profile. Therefore, even for some probiotic strains, mainly members of *Lactobacillus* and *Bifidobacterium*, that have a long history of use and are generally recognized as safe (GRAS status) [2], the FAO/WHO Working Group recommends characterization with several tests to assure safety [1]. These tests include determination of antibiotic resistance patterns, assessment for certain metabolic activities, tests for toxin production, determining if the strain under evaluation belongs to a species that is a known mammalian toxin producer, evaluation for hemolytic activity, determining if the strain belongs to a species with known hemolytic potential, monitoring side-effects during human studies, and post-marketing epidemiological surveillance of adverse incidents [1].

Probiotics have been demonstrated to be effective in a wide variety of conditions, including acute viral gastroenteritis, antibiotic-associated diarrhea, upper respiratory tract infections, atopic eczema, and some inflammatory conditions [3]. Probiotics have an ability to colonize the human intestinal lumen, multiply, and survive. Their effects on the host immunological system is not completely understood, but include blocking pathogenic bacterial effects and modulation of the host immune system by interacting with intestinal epithelial cells and antigen-presenting cells, thus initiating immunological responses [4]. Moreover, probiotics have been found to enhance innate immunity and modulate pathogen-induced inflammation via Toll-like receptor-regulated signaling pathways [5].

Taking into account that probiotics are live microorganisms and that these probiotics can influence the human immune system, several safety issues need to be addressed. These theoretical risks include the potential for transmigration causing bacteremia and sepsis, the potential for antibiotic resistance, and adverse effects influencing gastrointestinal physiology or various immunological functions [6, 7]. The aim of this review is to summarize the available data on the safety of probiotics for human use.

Adverse Events during Randomized Controlled Trails

While the number of randomized controlled trials (RCTs) evaluating the use of probiotics in different diseases has increased exponentially, not all have reported on whether and how adverse events were recorded during the study. A recently published report on the safety of probiotics by the US Agency for Healthcare Research and Quality (AHRQ) summarized all the available evidence on this often neglected issue [8]. The report evaluated the data concerning the safety of the use of probiotic organisms from 6 genera: *Lactobacillus, Bifidobacterium, Saccharomyces, Streptococcus, Enterococcus, and Bacillus* spp., alone, or in combinations. After a systematic review of 11,981 publications, of which 622 studies were included in the review, the authors found that only 387 studies reported the presence or absence of specific adverse events. The most important limitation of the report, as stated by the authors, is that most of the included RCTs were not designed to evaluate the safety of the studied probiotic strains. Based on reported adverse events, the RCTs showed no statistically significant increase in the relative risk (RR) for all adverse events (RR 1.00; 95% CI: 0.93, 1.07; p = 0.999) or serious adverse events (RR 1.06; 95% CI: 0.97, 1.16; p = 0.201) with short-term probiotic use in comparison to control groups that did not receive probiotics. The most commonly reported adverse events included gastrointestinal adverse events (diarrhea, vomiting, colic, and constipation) and infections caused by the administered strain. However, there was no evidence that participants using probiotic organisms experienced significantly more of these specific events in comparison to controls. The report concluded that there was a lack of systematic reporting of adverse events in probiotic intervention studies and that the available evidence failed to answer specific questions on the safety of probiotics with confidence. Moreover, although not stated in the limitations, included studies could not ensure that the various control groups had not been exposed to probiotics as these organisms are used in a wide range of foods. On the other hand, one can argue that probiotics are not drugs and that we need to take into consideration that traditional foods and food components are not studied in the same way as drugs. Thus, the lack of reported adverse events throughout the substantial number of studies presented in the AHRQ report favors the argument that the use of probiotics could be generally accepted as safe [9].

Translocation and Infection

The incidence of bacteremia attributable to specific probiotic strains remains very low despite the widespread use of probiotics in foods and dietary supplements [10]. However, in recent years, many species of the genera *Lactobacillus*, *Enterococcus,* and *Bifidobacterium* have been isolated from patients with different severe infections, including endocarditis and bloodstream infections [7, 11, 12]. This phenomenon could indicate a probiotic's ability to translocate from the human gut to the bloodstream [2]. A generally accepted theory is that in critically ill patients, translocation of luminal bacteria is linked to disruption of the normal commensal microbiota with overgrowth of potentially pathogenic strains, disruption of the mucosal barrier, and an impaired host immune response [13]. The potential vulnerability of the critically ill patient was emphasized by the PROPATRIA study [14]. In this study, 296 patients with predicted severe acute pancreatitis were allocated to receive either a probiotic mixture (four *Lactobacillus* strains and two *Bifidobacterium* strains) or placebo. The incidence of infectious complications was comparable between the groups; however, mortality rates were significantly higher in the probiotic group (16 vs. 6%) as was the incidence of bowel ischemia [14]. Following the results of the PROPATRIA trial, the same group of authors evaluated the effects of probiotics on enterocyte damage, intestinal permeability, and bacterial translocation [15]. In that study [15], they showed that probiotic prophylaxis reduces bacterial translocation. However, bacterial translocation was increased in patients with organ failure, indicating that while probiotics have beneficial effects in the moderately ill patient, they may negatively affect the critically ill [15]. In patients with severe pancreatitis, time of administration could also play an important role in adverse reactions; it seems that probiotics administered after the onset of acute pancreatitis may act as an extra oxidative burden in already critically affected patients [16]. In contrast, probiotics administered before an expected oxidative assault might result in an enhanced antioxidative capacity [16]. The importance of the timing of probiotic administration was proven by clinical RCTs in patients undergoing surgery. In studies which used probiotic treatment preoperatively, a significant reduction in postoperative bacterial infections was found [17, 18], whereas such benefits were not unanimously seen in studies which employed a postoperative treatment strategy [19–21].

Despite the above, the use of probiotics in critically ill patients seems to be promising, including the benefits shown in antibiotic-associated diarrhea, ventilator-associated pneumonia, and necrotizing enterocolitis [22]; however, in this group of patients, and the lessons learned from the PROPATRIA study, these concepts need to be carefully studied for each indication and each specific strain with meticulous safety monitoring [22].

As the usage of probiotics is on the rise, the number of reports of sepsis caused by probiotic strains is on the rise as well. Until the year 2000, we were able to identify 13 published cases of sepsis caused by probiotic strains, with a significant rise in the number of cases thereafter (table 1). Evident from table 1 is the observation that all patients had an underlying disease: they were either immunocompromised, severely ill, treated in intensive care units, had several comorbidities, or had central venous catheters. Several of these reports originated from pediatric patients [23–33], most of these instances featured *Lactobacillus* strains and *Saccharomyces* species, with *Bifidobacterium* strains being infrequent [30, 32]. From this evidence, it is clear that probiotic treatment in patients who are critically ill should be implemented with caution.

Transferability of Antibiotic Resistance

A major area of concern has been the potential for the transfer of antibiotic resistance between commensal and pathogenic bacteria in the gastrointestinal tract [6]. Moreover, the increased level of ingested probiotic bacteria has caused new speculations that these bacteria might also contribute to the overall reservoir of antibiotic resistance genes which could be transferred to the indigenous flora or even pathogens [34]. Many studies have selected and identified antibiotic-resistant species of *Lactobacilli* from various foods, including fermented milk products, cheese, and different meat products [35]. However, in these instances the resistance towards several antibiotics is intrinsic in nature, implying a low transmission potential compared to plasmid-mediated resistance [36]. There are several studies investigating horizontal (plasmid-mediated) transferability of antibiotic resistance; however, most of these were documented in an in vitro environment and thus cannot be readily translated to in vivo circumstances [37–39]. Only a limited number of studies have investigated antibiotic

Table 1. Published case reports on the probiotic strain caused sepsis; cases with history of probiotic intake +/– confirmed same probiotic strain from the product and bloodstream

Author	Age	Probiotics	Underlying disease	Treatment	Outcome
Perapoch et al. [23], 2000	3 months	*S. cerevisiae*	congenital cardiopathy	amphotericin B	recovery
Hennequin et al. [24], 2000	30 months	*S. boulardii*	cystic fibrosis, CVC	amphotericin B	recovery
	36 years		HIV, CVC	fluconazole	recovery
	47 years		esophageal carcinoma, CVC	fluconazole	recovery
	78 years		COPD, chronic renal disease	none	recovery
Rijnders et al. [51], 2000	74 years	*S. boulardii*	subarachnoidal hematoma, neurosurgery, CVC	fluconazole	death
Cesaro et al. [25], 2000	8 months	*S. boulardii*	acute myeloid leukemia, neutropenia, CVC	amphotericin B	recovery
Presterl et al. [52], 2001	23 years	*L. rhamnosus*	bicuspid aortic valve, diabetes insipidus	penicillin G	recovery
Lherm et al. [53], 2002	50 years	*S. boulardii*	cardiac arrest	fluconazole	death
	51 years		aortic surgery/cachexia	fluconazole	death
	50 years		ARDS/gastric ulcer	none	recovery
	82 years		acute resp. failure	none	recovery
	75 years		acute resp. failure/	fluconazole	recovery
	77 years		peritonitis/duodenal ulcer	amphotericin B	death
	71 years		hemorrhagic CVS	none	recovery
Lestin et al. [54], 2003	48 years	*S. cerevisiae*	diabetes, multiple comorbidities, immunocompromised	none	death
Riquelme et al. [55], 2003	42 years	*S. boulardii*	kidney-pancreas transplantation, immunosuppression	fluconazole	recovery
	41 years		HIV	amphotericin B	
Kniehl et al. [56], 2003	79 years	*B. cereus*	coronary bypass surgery	not declared	not declared
	65 years		coronary bypass surgery		
	55 years		not mentioned		
Lungarotti et al. [26], 2003	3 weeks	*S. boulardii*	prematurity	amphotericin B	recovery
Cherifi et al. [57], 2004	89 years	*S. cerevisiae*	anorexia nervosa, pseudomembranous colitis	fluconazole	death
Kunz et al. [27], 2004	36 weeks gestation	*Lactobacillus* GG	short gut syndrome	ampicillin	recovery
	34 weeks gestation		gastroschisis	ceftriaxone + ampicillin	recovery
Henry et al. [58], 2004	65 years	*S. cerevisiae*	oropharyngeal carcinoma	amphotericin B	recovery
Burkhardt et al. [59], 2005	19 years	*S. boulardii*	spastic tetraparesis, hydrocephalus, ventriculoperitoneal shunt	voriconazole	recovery
De Groote et al. [28], 2005	11 months	*L. rhamnosus* GG	short bowel syndrome, CVC	ampicillin + gentamicin	recovery
Land et al. [29], 2005	6 weeks	*L. rhamnosus* GG	heart surgery, CVC	penicillin G + gentamicin	recovery
	6 years		cerebral palsy, microcephaly, mental retardation, seizure, CVC	ampicillin	recovery
Munoz et al. [60], 2005	76 years	*S. cerevisiae*	diabetes, heart surgery	fluconazole	death
	72 years		heart surgery	none	death
	74 years		rheumatoid arthritis, heart surgery	fluconazole	death

Table 1. Continued

Author	Age	Probiotics	Underlying disease	Treatment	Outcome
Ledoux et al. [61], 2006	38 years	*L. acidophilus*	AIDS, CVC	clindamycin + gentamicin	recovery
Piechno et al. [62], 2007	61 years	*S. boulardii*	pyriform sinus cancer	voriconazole	recovery
Lolis et al. [63], 2008	56 years	*S. boulardii*	acute pulmonary edema	caspofungin	recovery
Tommasi et al. [64], 2008	66 years	*L. casei*	hypertension, diverticulosis, COPD	levofloxacin	recovery
Zein et al. [65], 2008	54 years	*L. rhamnosus*	diabetes type 2	amoxicillin	recovery
Conen et al. [66], 2009	38 years	*L. rhamnosus*	ulcerative colitis, neck abscess	imipenem + clinda-mycin + fluconazole	recovery
Ohishi et al. [30], 2010	2 days	*B. breve* BBG-01	omphalocele, surgery	ampicillin/ sulbactam + amikacin	recovery
Russo et al. [67], 2010	75 years	*L. casei*	aortic dissection, placement of a prosthesis	ampicillin	recovery
Kochan et al. [68], 2011	24 years	*L. rhamnosus*	aortic valve replacement	not declared	recovery
Jenke et al. [32], 2011	21 days	*B. longum* and the probiotic *infantis* strain	VLBW, premature	cefotaxime + vancomycin + metronidazole	recovery
Lee et al. [33], 2011	2 years	*Lactobacillus* spp.	Acute lymphoblastic leukemia	ceftriaxone	recovery
Stefanatou et al. [69], 2011	34 years	*S. cerevisiae*	thermal burns, CVC	none	death

ARDS = Acute respiratory distress syndrome; COPD = chronic obstructive pulmonary disease; CVC = central venous catheter; CVS = cerebrovascular stroke; HIV = human immunodeficiency virus; VLBW = very low birthweight.

resistance transfer from *Lactobacillus* involving conjugative plasmid pAMb1 encoding resistance to macrolide, lincosamide, and streptogramin B antibiotics in vivo; this phenomenon has been shown for *Lactobacillus reuteri* which was able to transfer pAMb1 to *Enterococcus faecalis* [40].

Moreover, wild-type *Lactobacillus plantarum* was able to horizontally transfer tetracycline and erythromycin resistance genes in vivo to *E. faecalis*, but in a lower fraction compared to an in vitro environment [41]. On the contrary, it has been shown that the same strain was able to transfer a small plasmid pLFE1 harboring the erythromycin resistance gene erm(B) to *E. faecalis* in vivo in higher frequency than in vitro, implying that the gastrointestinal tract may provide a more favorable environment for antibiotic resistance transfer [42]. Transferability of antibiotic resistance was not unanimously recognized; when tested in humans, *L. reuteri* failed to transfer the gene tet(W) to *Enterococci, Bifidobacteria,* and other *Lactobacilli* although the authors could not exclude a transfer at a lower rate to other components of the fecal bacterial

Hojsak · Shamir

population [43]. Although all of these studies indicate that antibiotic-resistant factors may be transferred from food-related bacteria to other, potentially pathogenic, species, its role in human health still remains controversial.

Safety of Long-Term Usage

The recently published systematic review by the European Society for Pediatric Gastroenterology, Hepatology and Nutrition (ESPGHAN) on the supplementation of infant formula with probiotics and/or prebiotics addressed the issue of safety by stating that the administration of probiotics in early infancy for a long period of time could have important implications. Nevertheless, the available scientific data suggest that the administration of evaluated probiotic-supplemented formula to healthy infants does not raise safety concerns [44]. Similar conclusions were reached by the American Academy of Pediatrics (AAP) on the supplementation of probiotics in infants [45]. Regarding long-term use, there are only a few RCTs which have addressed the long-term effects of probiotics and involved follow-up assessments of 1 year or more [46–50]. All of these controlled trials involved *Lactobacillus* strains, alone or in combination with *Bifidobacterium;* most of these studies commenced probiotic use in expectant mothers prenatally with probiotic supplementation continuing into the first year postpartum [46–48, 50]. The rate of adverse events was very low and a meta-analysis of these results showed a similar relative risk for adverse events in the probiotic treated and control groups (RR 0.76; 95% CI: 0.41, 1.39; p = 0.259) [8].

In summary, the long-term effects of probiotics are largely unknown, especially regarding the administration at the youngest age range. Large cohort studies with long-term follow-up are needed to fully understand the safety of probiotics among investigated populations.

Conclusion

Overall, there has been inadequate assessment and an absence of systematic reporting of adverse events in probiotic intervention studies. The evidence currently available from RCTs does not indicate an increased risk from the usage of any probiotic species in generally healthy individuals. However, special caution is needed in immunocompromised and severely ill patients. Moreover, when using any probiotic strain, premarketing assessment of safety should always be done and comprehensive postmarketing surveillance should be assured.

References

1 Joint FAO/WHO Working Group Report on Drafting Guidelines for the Evaluation of Probiotics in Food. London, April 30 and May 1, 2002.
2 Liong MT: Safety of probiotics: translocation and infection. Nutr Rev 2008;66:192–202.
3 Floch MH, Walker WA, Madsen K, Sanders ME, Macfarlane GT, Flint HJ, Dieleman LA, Ringel Y, Guandalini S, Kelly CP, Brandt LJ: Recommendations for probiotic use – 2011 update. J Clin Gastroenterol 2011;45(suppl):S168–S171.
4 Yan F, Polk DB: Probiotics and immune health. Curr Opin Gastroenterol 2011;27:496–501.
5 Vanderpool C, Yan F, Polk DB: Mechanisms of probiotic action: implications for therapeutic applications in inflammatory bowel diseases. Inflamm Bowel Dis 2008;14:1585–1596.
6 Snydman DR: The safety of probiotics. Clin Infect Dis 2008;46(Suppl 2):S104–S111, discussion S144–S151.
7 Sanders ME, Akkermans LM, Haller D, Hammerman C, Heimbach J, Hormannsperger G, Huys G, Levy DD, Lutgendorff F, Mack D, Phothirath P, Solano-Aguilar G, Vaughan E: Safety assessment of probiotics for human use. Gut Microbes 2010;1:164–185.
8 Hempel S NS, Ruelaz A, Wang Z, Miles JNV, Suttorp MJ, Johnsen B, Shanman R, Slusser W, Fu N, et al: Safety of probiotics to reduce risk and prevent or treat disease. Evidence-based practice center under contract No. 11–E007. Rockville (MD). Agency for Healthcare Research and Quality. 2011. www.ahrq.gov/clinic/tp/probiotictp.htm.
9 Wallace TC, MacKay D: The safety of probiotics: considerations following the 2011 U.S. Agency for Health Research and Quality report. J Nutr 2011;141:1923–1924.
10 Salminen MK, Tynkkynen S, Rautelin H, Saxelin M, Vaara M, Ruutu P, Sarna S, Valtonen V, Jarvinen A: *Lactobacillus bacteremia* during a rapid increase in probiotic use of *Lactobacillus rhamnosus* GG in Finland. Clin Infect Dis 2002;35:1155–1160.
11 Cannon JP, Lee TA, Bolanos JT, Danziger LH: Pathogenic relevance of Lactobacillus: a retrospective review of over 200 cases. Eur J Clin Microbiol Infect Dis 2005;24:31–40.
12 Boyle RJ, Robins-Browne RM, Tang ML: Probiotic use in clinical practice: what are the risks? Am J Clin Nutr 2006;83:1256–1264, quiz 1446–1447.
13 Wang ZT, Yao YM, Xiao GX, Sheng ZY: Risk factors of development of gut-derived bacterial translocation in thermally injured rats. World J Gastroenterol 2004;10:1619–1624.
14 Besselink MG, van Santvoort HC, Buskens E, Boermeester MA, van Goor H, Timmerman HM, Nieuwenhuijs VB, Bollen TL, van Ramshorst B, Witteman BJ, Rosman C, Ploeg RJ, Brink MA, Schaapherder AF, Dejong CH, Wahab PJ, van Laarhoven CJ, van der Harst E, van Eijck CH, Cuesta MA, Akkermans LM, Gooszen HG: Probiotic prophylaxis in predicted severe acute pancreatitis: a randomised, double-blind, placebo-controlled trial. Lancet 2008;371:651–659.
15 Besselink MG, van Santvoort HC, Renooij W, de Smet MB, Boermeester MA, Fischer K, Timmerman HM, Ahmed Ali U, Cirkel GA, Bollen TL, van Ramshorst B, Schaapherder AF, Witteman BJ, Ploeg RJ, van Goor H, van Laarhoven CJ, Tan AC, Brink MA, van der Harst E, Wahab PJ, van Eijck CH, Dejong CH, van Erpecum KJ, Akkermans LM, Gooszen HG: Intestinal barrier dysfunction in a randomized trial of a specific probiotic composition in acute pancreatitis. Ann Surg 2009;250:712–719.
16 Ammori BJ: Role of the gut in the course of severe acute pancreatitis. Pancreas 2003;26:122–129.
17 Rayes N, Seehofer D, Theruvath T, Mogl M, Langrehr JM, Nussler NC, Bengmark S, Neuhaus P: Effect of enteral nutrition and synbiotics on bacterial infection rates after pylorus-preserving pancreatoduodenectomy: a randomized, double-blind trial. Ann Surg 2007;246:36–41.
18 Nomura T, Tsuchiya Y, Nashimoto A, Yabusaki H, Takii Y, Nakagawa S, Sato N, Kanbayashi C, Tanaka O: Probiotics reduce infectious complications after pancreaticoduodenectomy. Hepatogastroenterology 2007;54:661–663.
19 Rayes N, Seehofer D, Hansen S, Boucsein K, Müller AR, Serke S, Bengmark S, Neuhaus P: Early enteral supply of Lactobacillus and fiber versus selective bowel decontamination: a controlled trial in liver transplant recipients. Transplantation 2002;74:123–127.
20 Rayes N, Hansen S, Seehofer D, Müller AR, Serke S, Bengmark S, Neuhaus P: Early enteral supply of fiber and Lactobacilli versus conventional nutrition: a controlled trial in patients with major abdominal surgery. Nutrition 2002;18:609–615.
21 Kanazawa H, Nagino M, Kamiya S, Komatsu S, Mayumi T, Takagi K, Asahara T, Nomoto K, Tanaka R, Nimura Y: Synbiotics reduce postoperative infectious complications: a randomized controlled trial in biliary cancer patients undergoing hepatectomy. Langenbecks Arch Surg 2005;390:104–113.
22 Morrow LE, Gogineni V, Malesker MA: Probiotics in the intensive care unit. Nutr Clin Pract 2012;27:235–241.

23 Perapoch J, Planes AM, Querol A, Lopez V, Martinez-Bendayan I, Tormo R, Fernandez F, Peguero G, Salcedo S: Fungemia with *Saccharomyces cerevisiae* in two newborns, only one of whom had been treated with ultra-levura. Eur J Clin Microbiol Infect Dis 2000;19:468–470.

24 Hennequin C, Kauffmann-Lacroix C, Jobert A, Viard JP, Ricour C, Jacquemin JL, Berche P: Possible role of catheters in *Saccharomyces boulardii* fungemia. Eur J Clin Microbiol Infect Dis 2000;19:16–20.

25 Cesaro S, Chinello P, Rossi L, Zanesco L: *Saccharomyces cerevisiae* fungemia in a neutropenic patient treated with *Saccharomyces boulardii*. Support Care Cancer 2000;8:504–505.

26 Lungarotti MS, Mezzetti D, Radicioni M: Methaemoglobinaemia with concurrent blood isolation of Saccharomyces and Candida. Arch Dis Child Fetal Neonatal Ed 2003;88:F446.

27 Kunz AN, Noel JM, Fairchok MP: Two cases of *Lactobacillus bacteremia* during probiotic treatment of short gut syndrome. J Pediatr Gastroenterol Nutr 2004;38:457–458.

28 De Groote MA, Frank DN, Dowell E, Glode MP, Pace NR: *Lactobacillus rhamnosus* GG bacteremia associated with probiotic use in a child with short gut syndrome. Pediatr Infect Dis J 2005;24:278–280.

29 Land MH, Rouster-Stevens K, Woods CR, Cannon ML, Cnota J, Shetty AK: Lactobacillus sepsis associated with probiotic therapy. Pediatrics 2005;115:178–181.

30 Ohishi A, Takahashi S, Ito Y, Ohishi Y, Tsukamoto K, Nanba Y, Ito N, Kakiuchi S, Saitoh A, Morotomi M, Nakamura T: Bifidobacterium septicemia associated with postoperative probiotic therapy in a neonate with omphalocele. J Pediatr 2010;156:679–681.

31 Robin F, Paillard C, Marchandin H, Demeocq F, Bonnet R, Hennequin C: *Lactobacillus rhamnosus* meningitis following recurrent episodes of bacteremia in a child undergoing allogeneic hematopoietic stem cell transplantation. J Clin Microbiol 2010;48:4317–4319.

32 Jenke A, Ruf EM, Hoppe T, Heldmann M, Wirth S: Bifidobacterium septicaemia in an extremely low-birthweight infant under probiotic therapy. Arch Dis Child Fetal Neonatal Ed 2012;97:F217–F218.

33 Lee AC, Siao-Ping Ong ND: Food-borne bacteremic illnesses in febrile neutropenic children. Hematol Rep 2011;3:e11.

34 Schjorring S, Krogfelt KA: Assessment of bacterial antibiotic resistance transfer in the gut. Int J Microbiol 2011;2011:312956.

35 Teale CJ: Antimicrobial resistance and the food chain. J Appl Microbiol 2002;92(suppl):85S–89S.

36 Tynkkynen S, Singh KV, Varmanen P: Vancomycin resistance factor of *Lactobacillus rhamnosus* GG in relation to enterococcal vancomycin resistance (van) genes. Int J Food Microbiol 1998;41:195–204.

37 Gevers D, Huys G, Swings J: In vitro conjugal transfer of tetracycline resistance from Lactobacillus isolates to other Gram-positive bacteria. FEMS Microbiol Lett 2003;225:125–130.

38 Klare I, Konstabel C, Werner G, Huys G, Vankerckhoven V, Kahlmeter G, Hildebrandt B, Muller-Bertling S, Witte W, Goossens H: Antimicrobial susceptibilities of Lactobacillus, Pediococcus and Lactococcus human isolates and cultures intended for probiotic or nutritional use. J Antimicrob Chemother 2007;59:900–912.

39 Ouoba LI, Lei V, Jensen LB: Resistance of potential probiotic lactic acid bacteria and bifidobacteria of African and European origin to antimicrobials: determination and transferability of the resistance genes to other bacteria. Int J Food Microbiol 2008;121:217–224.

40 Morelli L, Sarra PG, Bottazzi V: In vivo transfer of pAM beta 1 from *Lactobacillus reuteri* to *Enterococcus faecalis*. J Appl Bacteriol 1988;65:371–375.

41 Jacobsen L, Wilcks A, Hammer K, Huys G, Gevers D, Andersen SR: Horizontal transfer of tet(M) and erm(B) resistance plasmids from food strains of *Lactobacillus plantarum* to *Enterococcus faecalis* JH2–2 in the gastrointestinal tract of gnotobiotic rats. FEMS Microbiol Ecol 2007;59:158–166.

42 Feld L, Schjorring S, Hammer K, Licht TR, Danielsen M, Krogfelt K, Wilcks A: Selective pressure affects transfer and establishment of a *Lactobacillus plantarum* resistance plasmid in the gastrointestinal environment. J Antimicrob Chemother 2008;61:845–852.

43 Egervarn M, Lindmark H, Olsson J, Roos S: Transferability of a tetracycline resistance gene from probiotic *Lactobacillus reuteri* to bacteria in the gastrointestinal tract of humans. Antonie Van Leeuwenhoek 2010;97:189–200.

44 Braegger C, Chmielewska A, Decsi T, Kolacek S, Mihatsch W, Moreno L, Piescik M, Puntis J, Shamir R, Szajewska H, Turck D, van Goudoever J: Supplementation of infant formula with probiotics and/or prebiotics: a systematic review and comment by the ESPGHAN committee on nutrition. J Pediatr Gastroenterol Nutr 2011;52:238–250.

45 Thomas DW, Greer FR: Probiotics and prebiotics in pediatrics. Pediatrics 2010;126:1217–1231.

46 Abrahamsson TR, Jakobsson T, Bottcher MF, Fredrikson M, Jenmalm MC, Bjorksten B, Oldaeus G: Probiotics in prevention of IgE-associated eczema: a double-blind, randomized, placebo-controlled trial. J Allergy Clin Immunol 2007;119:1174–1180.

47 Kopp MV, Hennemuth I, Heinzmann A, Urbanek R: Randomized, double-blind, placebo-controlled trial of probiotics for primary prevention: no clinical effects of Lactobacillus GG supplementation. Pediatrics 2008;121:e850–e856.

48 Kuitunen M, Kukkonen K, Juntunen-Backman K, Korpela R, Poussa T, Tuure T, Haahtela T, Savilahti E: Probiotics prevent IgE-associated allergy until age 5 years in cesarean-delivered children but not in the total cohort. J Allergy Clin Immunol 2009;123:335–341.

49 Ljungberg M, Korpela R, Ilonen J, Ludvigsson J, Vaarala O: Probiotics for the prevention of beta cell autoimmunity in children at genetic risk of type 1 diabetes – the PRODIA study. Ann NY Acad Sci 2006;1079:360–364.

50 Niers L, Martin R, Rijkers G, Sengers F, Timmerman H, van Uden N, Smidt H, Kimpen J, Hoekstra M: The effects of selected probiotic strains on the development of eczema (the PandA study). Allergy 2009;64: 1349–1358.

51 Rijnders BJ, Van Wijngaerden E, Verwaest C, Peetermans WE: Saccharomyces fungemia complicating Saccharomyces boulardii treatment in a non-immunocompromised host. Intensive Care Med 2000; 26:825.

52 Presterl E, Kneifel W, Mayer HK, Zehetgruber M, Makristathis A, Graninger W: Endocarditis by Lactobacillus rhamnosus due to yogurt ingestion? Scand J Infect Dis 2001;33:710–714.

53 Lherm T, Monet C, Nougiere B, Soulier M, Larbi D, Le Gall C, Caen D, Malbrunot C: Seven cases of fungemia with Saccharomyces boulardii in critically ill patients. Intensive Care Med 2002;28:797–801.

54 Lestin F, Pertschy A, Rimek D: Fungemia after oral treatment with Saccharomyces boulardii in a patient with multiple comorbidities (in German). Dtsch Med Wochenschr 2003;128:2531–2533.

55 Riquelme AJ, Calvo MA, Guzman AM, Depix MS, Garcia P, Perez C, Arrese M, Labarca JA: Saccharomyces cerevisiae fungemia after Saccharomyces boulardii treatment in immunocompromised patients. J Clin Gastroenterol 2003;36:41–43.

56 Kniehl E, Becker A, Forster DH: Pseudo-outbreak of toxigenic Bacillus cereus isolated from stools of three patients with diarrhoea after oral administration of a probiotic medication. J Hosp Infect 2003;55:33–38.

57 Cherifi S, Robberecht J, Miendje Y: Saccharomyces cerevisiae fungemia in an elderly patient with Clostridium difficile colitis. Acta Clin Belg 2004;59:223–224.

58 Henry S, D'Hondt L, Andre M, Holemans X, Canon JL: Saccharomyces cerevisiae fungemia in a head and neck cancer patient: a case report and review of the literature. Acta Clin Belg 2004;59:220–222.

59 Burkhardt O, Kohnlein T, Pletz M, Welte T: Saccharomyces boulardii induced sepsis: successful therapy with voriconazole after treatment failure with fluconazole. Scand J Infect Dis 2005;37:69–72.

60 Munoz P, Bouza E, Cuenca-Estrella M, Eiros JM, Perez MJ, Sanchez-Somolinos M, Rincon C, Hortal J, Pelaez T: Saccharomyces cerevisiae fungemia: an emerging infectious disease. Clin Infect Dis 2005;40: 1625–1634.

61 Ledoux D, Labombardi VJ, Karter D: Lactobacillus acidophilus bacteraemia after use of a probiotic in a patient with AIDS and Hodgkin's disease. Int J STD AIDS 2006;17:280–282.

62 Piechno S, Seguin P, Gangneux JP: Saccharomyces boulardii fungal sepsis: beware of the yeast (in French). Can J Anaesth 2007;54:245–246.

63 Lolis N, Veldekis D, Moraitou H, Kanavaki S, Velegraki A, Triandafyllidis C, Tasioudis C, Pefanis A, Pneumatikos I: Saccharomyces boulardii fungaemia in an intensive care unit patient treated with caspofungin. Crit Care 2008;12:414.

64 Tommasi C, Equitani F, Masala M, Ballardini M, Favaro M, Meledandri M, Fontana C, Narciso P, Nicastri E: Diagnostic difficulties of Lactobacillus casei bacteraemia in immunocompetent patients: a case report. J Med Case Reports 2008;2:315.

65 Zein EF, Karaa S, Chemaly A, Saidi I, Daou-Chahine W, Rohban R: Lactobacillus rhamnosus septicemia in a diabetic patient associated with probiotic use: a case report (in French). Ann Biol Clin (Paris) 2008; 66:195–198.

66 Conen A, Zimmerer S, Trampuz A, Frei R, Battegay M, Elzi L: A pain in the neck: probiotics for ulcerative colitis. Ann Intern Med 2009;151:895–897.

67 Russo A, Angeletti S, Lorino G, Venditti C, Falcone M, Dicuonzo G, Venditti M: A case of Lactobacillus casei bacteraemia associated with aortic dissection: is there a link? New Microbiol 2010;33:175–178.

68 Kochan P, Chmielarczyk A, Szymaniak L, Brykczynski M, Galant K, Zych A, Pakosz K, Giedrys-Kalemba S, Lenouvel E, Heczko PB: Lactobacillus rhamnosus administration causes sepsis in a cardiosurgical patient – is the time right to revise probiotic safety guidelines? Clin Microbiol Infect 2011;17:1589–1592.

69 Stefanatou E, Kompoti M, Paridou A, Koutsodimitropoulos I, Giannopoulou P, Markou N, Kalofonou M, Trikka-Graphakos E, Tsidemiadou F: Probiotic sepsis due to Saccharomyces fungaemia in a critically ill burn patient. Mycoses 2011;54:e643–e646.

Iva Hojsak, MD, PhD
Referral Center for Pediatric Gastroenterology and Nutrition
Children's Hospital Zagreb, University of Zagreb School of Medicine
Klaićeva 16, HR–10000 Zagreb (Croatia)
E-Mail ivahojsak@gmail.com

Guarino A, Quigley EMM, Walker WA (eds): Probiotic Bacteria and Their Effect on Human Health and Well-Being.
World Rev Nutr Diet. Basel, Karger, 2013, vol 107, pp 171–177 (DOI: 10.1159/000345746)

Age-Related Functional Feeding: A Novel Tool to Improve the Quality of Life

S.D. Forssten[a] · H. Röytiö[a] · F. Ibrahim[b] · A.C. Ouwehand[a]

[a]Danisco Sweeteners, Active Nutrition, DuPont Nutrition and Health, Kantvik, Finland;
[b]School of Science, Technology, and Health, University Campus Suffolk, Ipswich, UK

Abstract

At different stages of life, probiotics can provide health benefits to the consumer. Certain benefits, such as reduction in allergy, are typical for infants. Modulation of immune function and maintenance of the intestinal microbiota are mainly beneficial during infancy/childhood and at old age. Although it may seem that in adulthood, in the prime of one's life, probiotics have fewer benefits to offer, there are still substantial health gains possible, i.e. for lactose intolerance and irritable bowel syndrome and reduction in the risk for antibiotic-associated diarrhea and respiratory tract infections. Similar benefits, but usually with a more pronounced effect, can be observed in the elderly. With the exception of the relief of lactose intolerance symptoms, the European Food Safety Authority (EFSA) has, to date, not approved any health claims for probiotics. While this may seem a contradiction to the content of the chapter, it can be explained by a difference in interpretation of what a health benefit is. According to the regulation that EFSA has to follow, a health benefit is a reduction in a disease risk factor; here, however, we discuss the clinical endpoints.

By definition, probiotics should convey a health benefit to the consumer. As typical ingredients of functional foods, probiotics mainly exert their benefits through maintenance of health by reducing disease risk. The mechanisms through which these benefits are yielded are not always completely understood, but are increasingly being elucidated.

At different ages, consumers can benefit from the consumption of probiotics. At young age they may aid in establishing a beneficial microbiota and facilitate the development of the immune system in an appropriate direction. Also at old age, when the immune system exhibits reduced activity, probiotics have been documented to modulate its activity, improving its ability in pathogen surveillance while at the same time reducing auto immune responses. At all ages, specific probiotics have been ob-

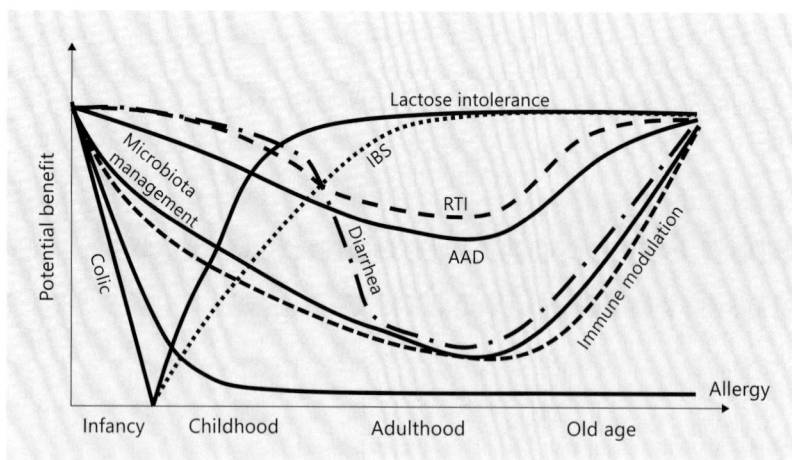

Fig. 1. Relative potential benefit of probiotics at different stages of life. IBS = Irritable bowel syndrome; RTI = respiratory tract infection.

served to reduce the incidence and duration of respiratory tract infections. In adults, especially when under stress or during extreme exercise, certain probiotics may reduce various inflammatory markers and the risk for infections.

Although various potential probiotic benefits are common for all ages, such as respiratory tract infections and antibiotic-associated diarrhea (AAD) risk, several benefits are directed at specific age populations (fig. 1). Despite this fact, no age-specific probiotics have been commercialized thus far. This may be related to the cost of developing a probiotic as it may not be commercially viable to do this only for a specific age group. It is also likely that this requires probiotics from genera other than those typically used as probiotics, i.e. lactobacilli and bifidobacteria, which may pose specific challenges.

Probiotics in Childhood

An increasing number of children in Western countries suffer from allergies, asthma, and atopic dermatitis (AD), which are regarded as a failure in the development of a balanced immune response. Dramatically increased prevalence implies that environmental factors have a role in the disease development, as the increase has been too fast for a shift in the genetic constitution to occur. The lack of microbial exposure early in life due to improved living conditions and changes in the diet are thought to lead to a 'dysbiosis' of the gut microbiota and skew the immune response against common and innocuous environmental antigens [1]. Since the intestinal microbiota has an active role in the modulation of immune responses, probiotic microbes have been suggested as protective or symptom-alleviating agents against

Forssten · Röytiö · Ibrahim · Ouwehand

allergic diseases. According to a meta-analysis, probiotic supplementation early in life and also prenatally (used by the mother during pregnancy) has a moderate role in preventing AD and IgE-associated AD [2]. Indeed, early feeding of *Lactobacillus rhamnosus* (HN001) has been shown to protect against eczema at 2 years of age by 50%, and this protective effect can be sustained at least until 4 years of age [3]. Similar results have been obtained with *L. rhamnosus* GG (LGG) against AD, and the key protective mechanism of LGG may be related to its ability to promote the growth of other beneficial microbial species in the gut, resulting in a more stable and functionally redundant gastrointestinal community of the infant [4]. Another meta-analysis concluded that especially the probiotic lactobacilli supplementation during pregnancy is effective in preventing AD in 2- to 7-year-old children [5]. Additionally, the use of LGG during pregnancy has also been shown to modify the excessive childhood weight gain of the offspring – another growing problem in the Westernized part of the world [6].

Interestingly, the dysbiosis of the infant microbiota has also been linked to excessive crying and fussing of the baby, a state that causes significant distress for the infant and the parents. Low proportions of *Bifidobacterium* and *Lactobacillus* in the feces have been correlated with total distress and the amount of crying [7]. This finding implies that probiotics could offer relief to colicky and also to less distressed infants.

Respiratory infections are common in children attending day care; the daily use of probiotic supplementation has reduced the episodes of fever, rhinorrhea, and cough as well as the cumulative duration of those symptoms. Also, the incidence of antibiotic prescriptions and the number of missed day care days has been reduced with *L. acidophilus* NCFM alone or in combination with *B. lactis* Bi-07 [8].

Furthermore, evidence is mounting that probiotics could have a role in preventing necrotizing enterocolitis, a more serious condition, in preterm infants. Several strains of probiotics are significantly able to reduce all-cause mortality and definite necrotizing enterocolitis without significant adverse effects in this vulnerable patient population [9]. Probiotics are also able to reduce the duration of acute watery diarrhea, particularly those due to rotavirus, by 1 day on average. This effect has been demonstrated at least for two probiotic strains, namely LGG and *Saccharomyces boulardii* [10].

Probiotics for Adults

A number of disease states of an adult may benefit from the use of probiotics, most notably lactose intolerance, diarrheal and other infectious bowel illnesses, and systemic infectious disorders.

Lactose intolerance can be ameliorated with the use of live yoghurt cultures *(L. delbrueckii* subsp. *bulgaricus* and *Streptococcus thermophilus)*. Yoghurt microbes (at least

10^8 CFU/g) aid the digestion of lactose in the yoghurt and decrease the symptoms in individuals with lactose maldigestion [11].

Probiotics have a positive effect on the incidence and duration of AAD. A recent meta-analysis [12] showed that the preventive effect of probiotics remained significant when grouped by probiotic species, population age group, relative duration of antibiotics and probiotics, study risk of bias, and probiotic administered. Another recent AAD study (unpubl. data, Ouwehand et al.) with a four-strain probiotic mix of *Lactobacillus* and *B. lactis* detected a significant decrease in both the incidence and duration of AAD.

Increasing evidence suggests that the intestinal microbiota has an important role in chronic mucosal inflammation occurring in inflammatory bowel disease, and a dysbiosis of the microbiota has been identified in these patients. Several mechanisms of action of probiotic products that may interfere with possible etiological factors in inflammatory bowel disease have been postulated. Positive effects have been detected for maintenance or induction of remission of inflammatory bowel disease; however, no clear evidence is available to support the use of probiotics in Crohn's disease [13]. Furthermore, the research on the effectiveness of probiotics for irritable bowel syndrome has a long way to go before any firm conclusions can be made. As the studies to date have used a variety of strains and formulas, it is difficult to make comparisons [14].

Probiotics and their potential role in enhancing immunity beyond the gastrointestinal tract have also been investigated. Probiotics may be useful for ameliorating upper respiratory tract infections in adults, as reduced length and severity of the infection have been shown with few *Lactobacillus* strains and *Bifidobacterium lactis* [15, 16].

Probiotics and their effects on preventing or treating AD have been studied in adult populations as well. *Lactobacillus salivarius* LS01 supplementation resulted in improvement in clinical parameters (SCORAD $p < 0.0001$ and DLQI $p = 0.021$) in adult AD patients. In addition, changes in the Th1/Th2 cytokine profiles could be detected; thus, the probiotic strain could have an important adjunctive therapy effect in the treatment of adult AD [17]. In another study, a probiotic mixture (*L. paracasei* Lpc-37, *L. acidophilus* 74–2 and *B. lactis* DGCC 420) lowered the genotoxicity of fecal water, which was elevated in AD patients [18].

Cardiovascular diseases are a major burden in the Western world and various preventive approaches have been investigated, including probiotics. In particular, *Lactobacillus* and *Bifidobacterium* species have been studied due to their anticholesterolemic potential. However, although several clinical trials have been conducted during the last two decades, no decisive outcome has been obtained thus far. The exact mechanism for cholesterol removal is poorly understood, although the 'bile salt hydrolase hypothesis' has been proposed to explain cholesterol-lowering effects of probiotics. Further studies are needed to determine whether the bile salt hydrolase activity of the probiotics strains is beneficial to the host [19].

Probiotics for Elderly

As discussed in the previous sections, various health benefits of probiotics have been documented for children and adults. Several studies have shown that the health benefits are even more pronounced at old age (>70 years). Some of the important findings will be briefly discussed.

The consumption of *L. johnsonii* La-1 by elderly volunteers normalized their response to endotoxin and modulated activation markers in blood phagocytes, suggesting potential to reduce low-grade chronic inflammation [20]. Intake of *B. longum* and *B. lactis* reduced serum IL-10 and increased anti-inflammatory TGF-β1 levels, thus providing a means of influencing inflammatory responses [21]. Moreover, increases of natural killer cell activity and phagocytosis was shown upon the intake of various probiotic strain such as *L. rhamnosus* HN001, *B. lactis* HN019, *L. delbrueckii* subsp. *bulgaricus* 8481, or *L. plantarum* CECT7315/7316 [22]. A number of probiotic strains have been observed to aid in the treatment of constipation, such as *L. casei* Shirota [23], *L. plantarum* CECT 7315/7316 [24], and *B. lactis* HN019 [25]. AAD is more common among elderly than younger adults: they require antibiotic treatment more often and are more prone to AAD. As discussed above, probiotics can be helpful in reducing the risk for and/or the duration of AAD. Elderly are also more prone to infective diarrhea. *L. casei* Shirota did not affect the incidence of norovirus infection in elderly, but it shortened the duration of fever after the disease onset to 1.5 days compared to 2.9 days in the control group [26]. In contrast, *E. coli* strain Nissle 1917 failed to reduce the prevalence of multidrug-resistant *E. coli* in elderly patients [27].

The effectiveness of influenza vaccination in preventing illness is lower in the elderly. Some probiotic strains have been reported to stimulate the response to influenza vaccination in elderly. *L. plantarum* CECT 7315/7316 increased the levels of influenza-specific IgA and IgG antibodies [28]. Similar effects have been reported for *L. casei* DN-114 001 [29]. *L. casei* DN-114 001 consumption by elderly was also associated with a decreased duration of upper respiratory tract infections such as rhinopharyngitis, but the cumulative number of chronic infectious diseases was not affected. However, a study showed that daily consumption of *L. casei* Shirota had no significant effect on the protection against respiratory symptoms [30].

Conclusions

As summarized above, specific probiotic strains have been documented to have various health benefits for particular age populations or the general population. Despite this, the European Food Safety Authority (EFSA) has to date (February 2013) not approved any health claim for probiotics. Although this is disappointing, it is explainable. The EU regulation that is applied by EFSA stipulates that for an ingredient to apply for a health claim, it has to reduce a marker for disease risk [31]; it should not cure, prevent, or mitigate a disease (as this would be a pharmaceutical application). Most of the data presented above deals with clinical endpoints and is therefore not appropriate for EFSA claim approval. Reduction of disease risk markers, however, requires the existence of generally accepted biomarkers or a thorough understanding of the mechanisms by which probiotics (and also other health ingredients) exert their effect. Unfortunately, this is currently not the case. Thus, although much and convincing data is available, it is a different kind of data that will need to be generated for regulatory approval. These regulations will impel further collaboration between the scientific community and the food industry to meet the new demands for the benefits of human health.

References

1 Fujimura KE, Slusher NA, Cabana MD, Lynch SV: Role of the gut microbiota in defining human health. Expert Rev Anti Infect Ther 2010;8:435–454.

2 Pelucchi C, Chatenoud L, Turati F, Galeone C, Moja L, Bach JF, La Vecchia C: Probiotics supplementation during pregnancy or infancy for the prevention of atopic dermatitis: a meta-analysis. Epidemiology 2012;23:402–414.

3 Wickens K, Black P, Stanley TV, Mitchell E, Barthow C, Fitzharris P, Purdie G, Crane J: A protective effect of *Lactobacillus rhamnosus* HN001 against eczema in the first 2 years of life persists to age 4 years. Clin Exp Allergy 2012;42:1071–1079.

4 Cox MJ, Huang YJ, Fujimura KE, Liu JT, McKean M, Boushey HA, Segal MR, Brodie EL, Cabana MD, Lynch SV: *Lactobacillus casei* abundance is associated with profound shifts in the infant gut microbiome. PLoS One 2010;5:e8745.

5 Doege K, Grajecki D, Zyriax BC, Detinkina E, Zu Eulenburg C, Buhling KJ: Impact of maternal supplementation with probiotics during pregnancy on atopic eczema in childhood – a meta-analysis. Br J Nutr 2012;107:1–6.

6 Luoto R, Kalliomäki M, Laitinen K, Isolauri E: The impact of perinatal probiotic intervention on the development of overweight and obesity: follow-up study from birth to 10 years. Int J Obes (Lond) 2010; 34:1531–1537.

7 Pärtty A, Kalliomäki M, Endo A, Salminen S, Isolauri E: Compositional development of Bifidobacterium and Lactobacillus microbiota is linked with crying and fussing in early infancy. PLoS One 2012; 7:e32495.

8 Leyer GJ, Li S, Mubasher ME, Reifer C, Ouwehand AC: Probiotic effects on cold and influenza-like symptom incidence and duration in children. Pediatrics 2009;124:e172–e179.

9 Deshpande GC, Rao SC, Keil AD, Patole SK: Evidence-based guidelines for use of probiotics in preterm neonates. BMC Med 2011;9:92.

10 Guandalini S: Probiotics for prevention and treatment of diarrhea. J Clin Gastroenterol 2011;45:S149–S153.

11 European Food Safety Authority: Scientific Opinion on the substantiation of health claims related to live yoghurt cultures and improved lactose digestion. EFSA J 2010;8:1763.

12 Videlock EJ, Cremonini F: Meta-analysis: probiotics in antibiotic-associated diarrhoea. Aliment Pharmacol Ther 2012;35:1355–1369.

13 Jonkers D, Penders J, Masclee A, Pierik M: Probiotics in the management of inflammatory bowel disease: a systematic review of intervention studies in adult patients. Drugs 2012;72:803–823.

14 Ki Cha B, Mun Jung S, Hwan Choi C, Song ID, Woong Lee H, et al: The effect of a multispecies probiotic mixture on the symptoms and fecal microbiota in diarrhea-dominant irritable bowel syndrome: a randomized, double-blind, placebo-controlled trial. J Clin Gastroenterol 2012;46:220–227.

15 Pregliasco F, Anselmi G, Fonte L, Giussani F, Schieppati S, Soletti L: A new chance of preventing winter diseases by the administration of synbiotic formulations. J Clin Gastroenterol 2008;42:S224–S233.

16 de Vrese M, Winkler P, Rautenberg P, Harder T, Noah C, et al: Probiotic bacteria reduced duration and severity but not the incidence of common cold episodes in a double blind, randomized, controlled trial. Vaccine 2006;24:6670–6674.

17 Drago L, Iemoli E, Rodighiero V, Nicola L, De Vecchi E, et al: Effects of *Lactobacillus salivarius* LS01 (DSM 22775) treatment on adult atopic dermatitis: a randomized placebo-controlled study. Int J Immunopathol Pharmacol 2011;24:1037–1048.

18 Roessler A, Forssten SD, Glei M, Ouwehand AC, Jahreis G: The effect of probiotics on faecal microbiota and genotoxic activity of faecal water in patients with atopic dermatitis: a randomized, placebo-controlled study. Clin Nutr 2011;31:22–29

19 Kumar M, Nagpal R, Kumar R, Hemalatha R, Verma V, et al: Cholesterol-lowering probiotics as potential biotherapeutics for metabolic diseases. Exp Diabetes Res 2012;2012:902917.

20 Schiffrin EJ, Parlesak A, Bode C, Bode JC, van't Hof MA, Grathwohl D, Guigoz Y: Probiotic yogurt in the elderly with intestinal bacterial overgrowth: endotoxaemia and innate immune functions. Br J Nutr 2009;101:961–966.

21 Ouwehand AC, Bergsma N, Parhiala R, Lahtinen S, Gueimonde M, Finne-Soveri H, Strandberg T, Pitkala K, Salminen S: Bifidobacterium microbiota and parameters of immune function in elderly subjects. FEMS Immunol Med Microbiol 2008;53:18–25.

22 Ibrahim F, Rovio S, Granlund L, Salminen S, Viitanen M, Ouwehand A: Probiotics and immunosenescence: cheese as carrier. FEMS Immunol Med Microbiol 2010;59:53–59.

23 Cassani E, Privitera G, Pezzoli G, Pusani C, Madio C, Iorio L, Barichella M: Use of probiotics for the treatment of constipation in Parkinson's disease patients. Minerva Gastroenterol Dietol 2011;57:117–121.

24 Bosch Gallego M, Espadaler Mazo J, Mendez Sanchez M, Perez Carre M, Farran Codina A, Audivert Brugue S, Bonachera Sierra MA, Cune Castellana J: Consumption of the probiotic Lactobacillus plantarum CECT 7315/7316 improves general health in the elderly subjects. Nutr Hosp 2011;26:642–645.

25 Waller PA, Gopal PK, Leyer GJ, Ouwehand AC, Reifer C, et al: Dose-response effect of Bifidobacterium lactis HN019 on whole gut transit time and functional gastrointestinal symptoms in adults. Scand J Gastroenterol 2011;46:1057–1064.

26 Yamada T, Nagata S, Kondo S, Bian L, Wang C, Asahara T, Ohta T, Nomoto K, Yamashiro Y: Effect of continuous probiotic fermented milk intake containing Lactobacillus casei strain Shirota on fever in mass infectious gastroenteritis rest home outbreak. Kansenshogaku Zasshi 2009;83:31–35.

27 Tannock GW, Tiong IS, Priest P, Munro K, Taylor C, Richardson A, Schultz M: Testing probiotic strain Escherichia coli Nissle 1917 (Mutaflor) for its ability to reduce carriage of multidrug-resistant E. coli by elderly residents in long-term care facilities. J Med Microbiol 2011;60:366–370.

28 Bosch M, Mendez M, Perez M, Farran A, Fuentes MC, Cune J: Lactobacillus plantarum CECT7315 and CECT7316 stimulate immunoglobulin production after influenza vaccination in elderly. Nutr Hosp 2012;27:504–509.

29 Boge T, Remigy M, Vaudaine S, Tanguy J, Bourdet-Sicard R, van der Werf S: A probiotic fermented dairy drink improves antibody response to influenza vaccination in the elderly in two randomised controlled trials. Vaccine 2009;27:5677–5684.

30 Van Puyenbroeck K, Hens N, Coenen S, Michiels B, Beunckens C, Molenberghs G, Van Royen P, Verhoeven V: Efficacy of daily intake of Lactobacillus casei Shirota on respiratory symptoms and influenza vaccination immune response: a randomized, double-blind, placebo-controlled trial in healthy elderly nursing home residents. Am J Clin Nutr 2012;95:1165–1171.

31 European Food Safety Authority: Guidance on the scientific requirements for health claims related to gut and immune function. EFSA J 2011;9:1984.

Arthur C. Ouwehand (PhD), Research Manager
Active Nutrition, DuPont Nutrition and Health
Sokeritehtaantie 20
FIN–02460 Kantvik (Finland)
E-Mail arthur.ouwehand@danisco.com

Guarino A, Quigley EMM, Walker WA (eds): Probiotic Bacteria and Their Effect on Human Health and Well-Being.
World Rev Nutr Diet. Basel, Karger, 2013, vol 107, pp 178–185 (DOI: 10.1159/000345747)

Use of Microbes to Fight Microbes

Colin Hill

Alimentary Pharmabiotic Centre and Department of Microbiology, University College Cork,
Cork, Ireland

Abstract

Microbes face a constant challenge to survive in any given ecological niche. Each microbial cell must compete with friends and foes alike for usually scarce resources (macro- and micronutrients), be able to react quickly to changing external conditions, and be able to rapidly respond to threats from existing and newly arrived competitors. This competition is particularly acute on body surfaces and reaches its apex in the gut, which contains what is probably the densest, most highly populated ecological niche on Earth. It is no wonder then that most microbes which occupy these niches have developed sophisticated strategies to promote their survival under these challenging conditions. These include strategies aimed at the destruction of rivals (and very often the most closely related strains or species can represent the greatest competitive challenge for any given niche), the expression of structures and strategies designed to assist a microbe to occupy selected sites within a given niche, tactics such as motility, and strategies aimed at outcompeting other microbes for scarce resources. It is tempting to consider that we could take advantage of these strategies to design therapies which would use harmless commensal bacteria to combat pathogenic bacteria. There are many instances where probiotic bacteria have been shown to play a role in limiting or even preventing infection. This chapter will consider some of the mechanisms involved in this phenomenon, and how we can exploit microbial competition to improve health status in humans.

> Tho' Nature, red in tooth and claw
> In Memorium by *Alfred Lord Tennyson*

Competition is the driving force of evolution, and any organism which survives in a particular ecological niche must be able to compete effectively with its neighbors. For single-celled microbes, survival is particularly precarious and, thus, mechanisms of competition are highly tuned. For example, in the densely colonized mammalian gut, cells must deal with physicochemical challenges such as bile, compete for limited macro- and micronutrients, and fight for limited opportunities to establish a foothold within a very dynamic environment (e.g. bind to a receptor on the epithelium). Some of these strategies are clearly understood, but the precise mechanisms by which some species maintain a competitive advantage are hampered for many gut microorganisms by the fact that they are largely unculturable with classical microbiological methodologies. While this limits our understanding of many of the most abundant species in the gut, studies performed on the cultivatable fraction have provided a good understanding of many of the mechanisms involved in survival in the gut. Ideally it should prove possible to exploit the strategies employed by bacteria in any particular niche to deliberately prevent or limit infection of the same niche by an invading pathogen. Of course, we must bear in mind that the same ecological forces have driven pathogen evolution to allow them to compete with the existing microbiota at their target site, so it may well be that one would have to amplify the level of competition to prevent infection. For example, the delivery of probiotics in high numbers may temporarily tilt the balance of competition in favor of the commensals, and thereby prevent or ameliorate an infection.

One can easily envisage at least three mechanisms by which a commensal or probiotic bacteria could compete with a potential pathogen at a particular body site (fig. 1). Firstly, one microbe can directly affect another, either through direct antagonism or by interfering with the physiology of a target organism (e.g. downregulating gene expression in another bacterium). Another mechanism could be competition for binding sites on body surfaces (or for essential nutrients), while another mechanism could involve immunomodulation in which a harmless organism stimulates an immune response which renders the environment less hospitable to a competing pathogen. This chapter is not intended to be an exhaustive review of the research in this field, but will present some examples of where probiotics have been shown to prevent or limit infectious events (where microbes fight microbes), and describe some of the mechanisms responsible. A short description of how effective therapies may be developed based on our accumulating knowledge of how bacteria compete in the gut will also be presented.

Microbes Fighting Microbes: Direct Antagonism

Many gut microbes have the ability to produce antimicrobial compounds, including antimicrobial peptides termed bacteriocins [1]. Screening programs have identified many bacteriocin producers in the gut, while metagenomics and molecular analysis have revealed the coding capacity for bacteriocin production in many as yet uncultured organisms [2]. The prevalence of this trait suggests that it performs a valuable role in microbe:microbe competition and that it could represent a useful mechanism through which one could tilt the balance against an invading pathogen. A recent review assessed the roles of bacteriocins as probiotic traits [3]. One study which is relevant in this regard used a five-strain probiotic mixture (4 lactobacilli and a *Pediococ-*

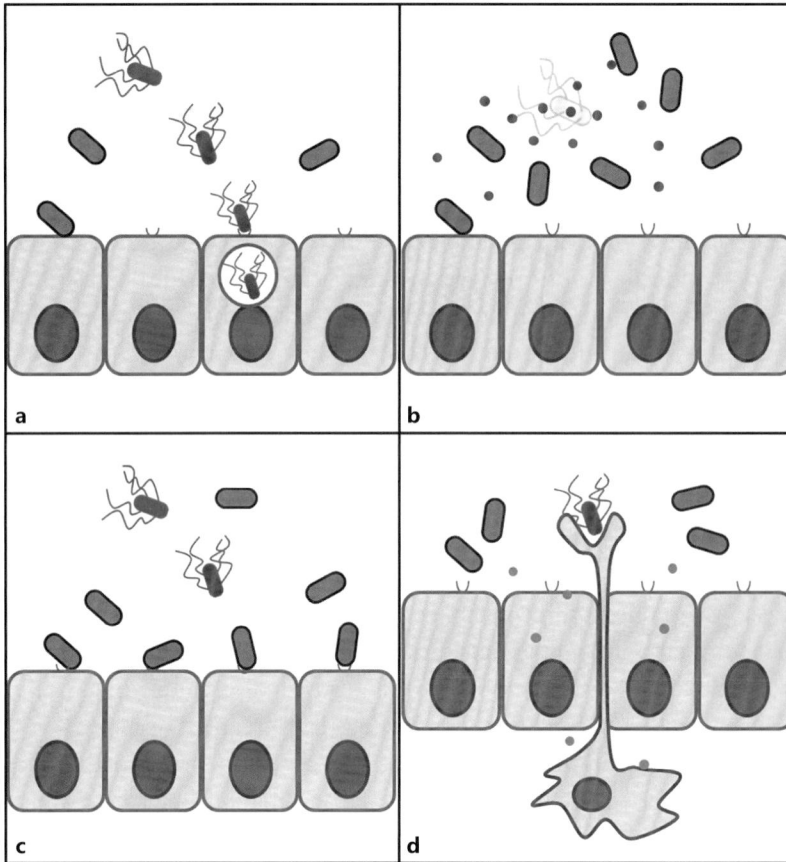

Fig. 1. Mechanisms by which microbes can prevent infection. **a** In the absence of sufficient competition, the pathogen gains access to the epithelium and invades the host. **b** The commensal strain produces metabolites which can directly impact on the viability of the pathogen or on the ability of the pathogen to appropriately express virulence factors. **c** The commensal cells inhabit the appropriate binding sites, preventing pathogen access to the epithelium. **d** The commensal signals the immune system in such a way as to render the pathogen less likely to invade.

cus) to protect pigs against infection with the enteropathogen *Salmonella enterica* [4]. The mixture was very successful in limiting infection, even completely preventing any symptoms in some instances. Interestingly, the single bacteriocin-producing strain was shown to dominate the other four nonproducers in the ileum digesta and mucosa (but not in feces). This implies that the ability to produce a bacteriocin is an important colonization factor and likely important in preventing the pathogen from mounting a successful infection. However, the specific role of the bacteriocin in controlling the *Salmonella* infection was not determined in this study.

There are many other examples of bacteriocin-producing strains linked to anti-infective phenomena. A strain of *Pediococcus acidilactici* MM33 which produces a bacteriocin, pediocin PA-1/AcH, could control vancomycin-resistant enterococci in mice, whereas a bacteriocin-nonproducing derivative had no effect [5]. Another elegant study demonstrated that a *Streptococcus mutans* strain producing mutacin 1140 can control plaque-forming bacteria [6]. This has been developed into a product concept by Oragenics termed SMaRT Replacement Therapy™ (http://www.oragenics.com/). It is noteworthy that the bacteriocin-producing strain could be recovered from the oral cavity of several individuals 14 years after a single inoculation [6]. *Streptococcus salivarius* K12 is another bacteriocin producer which has been shown to control bacteria involved in plaque formation and halitosis [7, 8]. Other bacteriocin-producing *S. salivarius* strains have also exhibited promising results against pharyngitis associated with *Streptococcus pyogenes* [9]. The ability of bacteriocin-producing strains to prevent vaginosis has also been demonstrated [10], albeit no direct role for the bacteriocin has as yet been proven.

More direct evidence of the role of in situ bacteriocin production as a mechanism for preventing infection was provided in a murine model of listeriosis. In this study a bacteriocin-producing strain, *Lactobacillus salivarius* UCC118, was shown to prevent infection of mice with *Listeria monocytogenes* [11]. The effect was dramatic in that mice prefed the UCC118 before infection survived, whereas those fed placebo all died. Two lines of evidence confirm the role of the bacteriocin in this protective phenomenon. Firstly, a bacteriocin-negative version of UCC118 which is identical in every other respect failed to protect mice against the *Listeria*, while a strain of *Listeria* rendered immune to the bacteriocin by cloning the immunity determinant from the producer strain was able to overcome the UCC118 protective effect and was lethal to the mice. This unequivocally demonstrates the potential for bacteriocin-producing strains to act prophylactically against infection.

These observations on the potential of bacteriocins to prevent or treat infection can obviously be extended clinically by isolating bacteriocins and using them either prophylactically or therapeutically, and much research has been done in this area. In one example, a bacteriocin was isolated from a human isolate of *Bacillus thuringiensis* which produces a narrow spectrum bacteriocin, thuricin CD [12]. This bacteriocin is highly active against *Clostridium difficile,* and was shown to specifically reduce *C. difficile* populations while having little impact on the gut microbiota in a human model of the distal colon [13].

Another form of microbe:microbe interaction in the gut is the interplay between bacteria and bacteriophages. Bacteriophages (phages) are bacterial viruses which can be carried within bacterial genomes (temperate or lysogenic phage) or exist as extracellular predators (lytic phages). Bacteriophages are likely to play a significant role in shaping microbial population structures in the human body, including the gut. By identifying and characterizing these phages we can gain a better understanding and control of population structure, and we can also use phages (or their lytic enzymes) as pharmabiotics in medicine and in food safety and preservation. Bacteriophages have been explored as potential therapeutic agents for many decades, particularly in the East, but Western medicine has relied on antibiotics rather than phages for control of infectious agents. Nonetheless, there is significant evidence that phages can work as effective therapeutics. In addition, the FDA has recently approved certain phage cocktails for use in food to prevent foodborne illnesses such as listeriosis. As an example of the experimental approach, a recent study used newly identified bacteriophages to control *Pseudomonas aeruginosa* in a mouse model of lung infection and in biofilms growing on ex vivo lung cells derived from a patient with cystic fibrosis [14]. The results were promising in that the mice treated with phages were rapidly cleared of the infection, and it was also notable that the phages were capable of eliminating bacteria in a mature biofilm formed on lung cells.

Microbes Fighting Microbes: Competition for Binding Sites and Nutrients

Microbes compete for limited sites in the small and large intestine, and there is a lot of evidence supporting the ability of probiotics to block access of pathogens to cell line models. It is a reasonable assumption that this phenomenon can be translated to the gut. For example, it has been shown that a well-known commercial probiotic, *Lactobacillus rhamnosus* GG, produces pili which are very similar to those expressed on the surface of vancomycin-resistant enterococci [15]. It is interesting to note an earlier paper reported GG could successfully treat gastrointestinal carriage of vancomycin-resistant enterococci in renal patients [16]. More recently, it has been shown that *Bifidobacterium* spp. possess TadIV pili (tight adherence pili) which are only expressed in vivo and which promote colonization of the murine gastrointestinal tract [17]. Before this discovery, TadIV pili were only associated with pathogens, such as *Yersinia enterocolitica* [18].

It has also been shown in an elegant study [19] that germ-free mice can be infected with *Citrobacter rodentium* and are unable to eliminate the pathogen. The study clearly demonstrated that the ability of subsequently introduced commensals to compete with *C. rodentium* was influenced by the ability of pathogens and commensals to utilize similar carbohydrates. They concluded that colonization by pathogens is a function of bacterial virulence gene expression and is affected by competition with commensals competing for similar growth substrates.

Microbes Fighting Microbes: Immunomodulation

Microbes occupying niches on the mammalian body are likely to come into contact with the immune system. This offers the potential for microbial interventions directed at stimulating a more aggressive response against pathogenic microbes. Of course the mucosal immune system is both complex and highly regulated; therefore, most examples of protective effects in this area have not resolved the precise mechanistic basis of microbes fighting microbes by proxy – through the immune system. Many

Fig. 2. Microbes fighting microbes in the mammary gland. The effect of infusing live *L. lactis* into the teat of an animal suffering mastitis is shown in the quality of the milk collected before (left) and after (right) treatment.

probiotic strains are marketed as having an immunomodulatory effect, but most of these effects have been elucidated in model systems such as immortalized cell lines. However, several lines of evidence suggest that immunomodulation is a valid mechanism through which commensal bacteria can prevent or limit infection.

One somewhat unusual example of an immunomodulatory mechanism involves bovine mastitis. This is an infection of the mammary gland and is one of the most costly animal diseases worldwide. Most cases are treated with traditional antibiotics. In a recent study, we used a probiotic approach to address this aggressive animal disease [20]. In brief, animals with mild or severe mastitis were randomly assigned to receive standard antibiotic therapy or to be infused with a harmless strain of *Lactococcus lactis* (fig. 2). The outcome suggested that the 'probiotic' treatment was as effective as the standard therapy in treating mastitis, with no statistically significant difference between the overall cure rates: 64% for the probiotic treatment and 72% for the antibiotic treatment. A follow-up investigation in uninfected animals revealed a significant increase in levels of polymorphonuclear leukocytes and lamina propria leukocytes in the mammary gland of the animals given the probiotic, providing a likely explanation for the protective effect [21]. In addition, increased acute phase proteins (haptoglobin and serum amyloid A) were observed in the milk in response to probiotic treatment.

Conclusions

The concept of using microbes to fight microbes is likely to become a very important strategy in infection control. It is obvious from germ-free animals that the infectious dose for many pathogens is significantly lower in the absence of competition from a commensal microbiota. It is a logical extension of this observation to seek to improve the barrier to infection provided by commensals through specific microbial therapies. These may take the form of prebiotic or probiotic interventions. One of the advantages of such approaches is the potential lack of side effects, while another benefit is that such strategies are unlikely to lead to the same resistance problems as has been observed to be associated with classical antibiotic therapies. We are likely to see many more examples of microbes fighting microbes being used as another weapon for clinicians in the fight against infections.

References

1 Cotter P, Hill C, Ross RP: Bacteriocins: developing innate immunity for food. Nat Rev Microbiol 2005; 3:777–788.

2 de Jong A, van Heel AJ, Kok J, Kuipers OP: BAGEL2: mining for bacteriocins in genomic data. Nucleic Acids Res 2010;38:647–651.

3 Dobson A, Cotter PD, Ross RP, Hill C: Bacteriocin production as a probiotic trait. Appl Environ Microbiol 2012;78:1–6.

4 Casey PG, Gardiner GE, Casey G, Bradshaw B, Lawlor PG, Lynch PB, Leonard FC, Stanton C, Ross RP, Hill C, Fitzgerald GF: A five-strain probiotic combination reduces pathogen shedding and alleviates disease signs in pigs challenged with *Salmonella typhimurium*. Appl Environ Microbiol 2007;73:1858–1863.

5 Millette M, Cornut G, Dupont C, Shareck F, Archambault D, Lacroix M: Capacity of human nisin- and pediocin-producing lactic acid bacteria to reduce intestinal colonization by vancomycin-resistant enterococci. Appl Environ Microbiol 2008;74:1997–2003.

6 Hillman J: Genetically modified *Streptococcus mutans* for the prevention of dental caries. Antonie Van Leeuwenhoek 2002;82:361–366.

7 Balakrishnan M, Simmonds RS, Carne A, Tagg JR: *Streptococcus mutans* strain N produces a novel low molecular mass non-lantibiotic bacteriocin. FEMS Microbiol Lett 2000;183:165–169.

8 Burton JP, Chilcott CN, Moore CJ, Speiser G, Tagg JR: A preliminary study of the effect of probiotic *Streptococcus salivarius* K12 on oral malodour parameters. J Appl Microbiol 2006;100:754–764.

9 Tagg JR: Prevention of streptococcal pharyngitis by anti-*Streptococcus pyogenes* bacteriocin-like inhibitory substances (BLIS) produced by *Streptococcus salivarius*. Indian J Med Res 2004;119(suppl):13–16.

10 Dover SE, Aroutcheva AA, Faro S, Chikindas ML: Natural antimicrobials and their role in vaginal health: a short review. Int J Probiotics Prebiotics 2008;3:219–230.

11 Corr SC, Li Y, Riedel CU, O'Toole PW, Hill C, Gahan CG: Bacteriocin production as a mechanism for the anti-infective activity of *Lactobacillus salivarius* UCC118. Proc Natl Acad Sci USA 2007;104:7617–7621.

12 Rea M, Sit CS, Clayton E, O'Connor PM, Whittal RM, Zheng J, Vederas JC, Ross RP, Hill C: Thuricin CD, a post-translationally modified bacteriocin with a narrow spectrum of activity against *Clostridium difficile*. Proc Natl Acad Sci USA 2010;107:9352–9358.

13 Rea M, Dobson A, O'Sullivan O, Crispie F, Fouhy F, Cotter PD, Shanahan F, Kiely B, Hill C, Ross RP: Effect of broad and narrow spectrum antimicrobials on *Clostridium difficile* and microbial diversity in a model of the distal colon. Proc Natl Acad Sci USA 2011;108:4639–4644.

14 Alemayehu D, Casey P, McAuliffe O, Guinane CM, Martin JG, Shanahan F, Coffey A, Ross RP, Hill C: Bacteriophages ΦMR299–2 and ΦNH-4 can eliminate *Pseudomonas aeruginosa* in the murine lung and on cystic fibrosis lung airway cells. MBio 2012; 3:e00029–12.

15 Kankainen M, Paulin L, Tynkkynen S, von Ossowski I, Reunanen J, Partanen P, Satokari R, Vesterlund S, Hendrickx AP, Lebeer S, De Keersmaecker SC, Vanderleyden J, Hämäläinen T, Laukkanen S, Salovuori N, Ritari J, Alatalo E, Korpela R, Mattila-Sandholm T, Lassig A, Hatakka K, Kinnunen KT, Karjalainen H, Saxelin M, Laakso K, Surakka A, Palva A, Salusjärvi T, Auvinen P, de Vos WM: Comparative genomic analysis of *Lactobacillus rhamnosus* GG reveals pili encoding a human-mucus binding protein. Proc Natl Acad Sci USA 2009;106: 17193–17198.

16 Manley KJ, Fraenkel MB, Mayall BC, Power DA: Probiotic treatment of vancomycin-resistant enterococci: a randomised controlled trial. Med J Aust 2007;186:454–457.

17 Motherway MO, Zomer A, Leahy SC, Reunanen J, Bottacini F, Claesson MJ, O'Brien F, Flynn K, Casey PG, Munoz JAM, Kearney B, Houston AM, O'Mahony, C, Higgins DG, Shanahan F, Palva A, de Vos WM, Fitzgerald GF, Ventura M, O'Toole PW, van Sinderen D: Functional genome analysis of *Bifidobacterium breve* UCC2003 reveals type IVb tight adherence (Tad) pili as an essential and conserved host-colonization factor. Proc Natl Acad Sci USA 2011;108:11217–11222.

18 Schilling J, Wagner K, Seekircher S, Greune L, Humberg V, Schmidt MA, Heusipp G: Transcriptional activation of the tad type IVb pilus operon by PypB in *Yersinia enterocolitica*. J Bacteriol 2010;192:3809–3821.

19 Kamada N, Kim Y-G, Sham HP, Vallance BA, Puente JL, Martens EC, Núñez G: Regulated virulence controls the ability of a pathogen to compete with the gut microbiota. Science 2012;336:1325–1329.

20 Klosterman KF, Crispie F, Flynn J, Ross RP, Hill C, Meaney W: Intramammary infusion of a live culture of *Lactococcus lactis* for treatment of bovine mastitis: comparison to antibiotic treatment in field trials. J Dairy Res 2008;75:365–373.

21 Crispie F, Alonso-Gomez M, O'Loughlin C, Klostermann K, Flynn J, Arkins S, Meaney W, Ross RP, Hill C: Intramammary infusion of a live culture for treatment of bovine mastitis: effect of live lactococci on the mammary immune response. J Dairy Res 2008;75:374–384.

Colin Hill
Alimentary Pharmabiotic Centre and Department of Microbiology
University College Cork
Cork (Ireland)
E-Mail c.hill@ucc.ie

Guarino A, Quigley EMM, Walker WA (eds): Probiotic Bacteria and Their Effect on Human Health and Well-Being.
World Rev Nutr Diet. Basel, Karger, 2013, vol 107, pp 186–196 (DOI: 10.1159/000346491)

What Is the Future for Therapies Derived from the Microbiome (Pharmabiotics)?

Ger T. Rijkers[a] · Linda Mulder[b] · Frans M. Rombouts[c] ·
Louis M.A. Akkermans[d]

[a] Biomedical and Life Sciences, Utrecht University, Utrecht and Roosevelt Academy, Middelburg,
[b] Winclove, Amsterdam, [c] Food Microbiology, Wageningen University, Wageningen, [d] Gastrointestinal
Physiology, Utrecht University, Utrecht and University Medical Center Utrecht, Utrecht, The Netherlands

Abstract

The personal gut microbiota is characterized by species composition, enterotype, and bacterial gene counts. Gut microbiota can be viewed as a complex microecosystem. Regulation of the diversity and stability of the gut microbiota is therefore critical for the development of future therapies. The areas with high potential for personalized management of gut microbiota are obesity and the metabolic syndrome, prevention and control of (recurrent) infections, immune-mediated disorders, and the gut-brain axis. A true and deeper understanding of the interaction between the microbiota and the host, as well as a better matching of probiotic and prebiotic mechanisms with clinical indications will be required for successful future implementation of these therapies.

Copyright © 2013 S. Karger AG, Basel

Because of the coevolution of man and microbes, the human intestinal tract is colonized by thousands of species of bacteria. Gut-borne microbes outnumber the total number of body tissue cells by a factor of ten. Recent metagenomics analysis of the human gut microbiota has revealed the presence of some 3.3 million genes, as compared to the mere 23,000 genes present in the cells of the human body tissues [1]. On average, each individual has approximately 540,000 of the initial 3.3 million genes. Similarity between individuals is reflected in the core metagenome genes: approximately 50% of an individual's genes are shared by at least 50% of individuals of the cohort. On the other hand, individuality is determined by rare genes: the genes shared by less than 20% of individuals encompass 2.4 million genes. Thus, we (our gut microbiota) are all rather similar but not identical. Based on the sequenced metagenomes, individuals can be grouped into three robust clusters (referred to as entero-types). Each enterotype is characterized by a different bacterial ecosystem, with a high abundance of *Bacteroides*, *Prevotella*, or *Ruminococcus* [2, 3].

People may differ by species composition and enterotype, as well as by gut bacterial gene counts. The human intestinal microbiota thus shares large similarities as well as differences that permit stratification, with potential applications in personalized or digitized medicine and nutrition.

Me, myself, I by Joan Armatrading was a signature song of the 1980s. We now realize that man lives in intimate association with its gut microbiota, and not alone. The cover story of *The Economist* (August 18, 2012) on gut microbiota therefore was appropriately entitled '*Me, myself, us*'.

Perturbation of the intestinal microbiota may lead to chronic diseases such as autoimmune diseases, colon cancers, gastric ulcers, cardiovascular disease, and obesity. Restoration of the gut microbiota may be difficult to accomplish, but the use of pro- and prebiotics has led to promising results in a large number of well-designed (clinical) studies (reviewed elsewhere in this volume). Microbiomics has spurred a dramatic increase in scientific, industrial, and public interest in probiotics and prebiotics as possible agents for gut microbiota management and control. Genomics and bioinformatics tools may allow us to establish mechanistic relationships between gut microbiota and the health status of the individual, hopefully providing perspectives for personalized gut microbiota management.

The above themes were addressed in an international workshop (www.gutmicrobiota.org; September 13–15, 2012; Maastricht, The Netherlands) from various different angles ranging from transitions in ecosystems to microbe-microbe and microbe-host interactions. Basic scientists (microbiology, immunology, systems biology) were teamed with clinical and nutritional specialists to pave the road to personalized gut microbiota management.

Gut microbiota can be viewed as a complex microecosystem. The work of the group of Marten Scheffer has shown that the stability of complex macroecosystems such as rainforests or lakes is maintained by common regulatory mechanisms which can be mathematically approached [4]. The mathematical model predicting loss of stability was used by Salvador Dali in his last painting, *The Swallow's Tail*, in 1983 (fig. 1). By using these models, it can also be predicted under which circumstances the system will lose its resilience, even when it reaches the tipping point [5, 6]. The gut microbiota may be an ecosystem to which these same rules apply. Thus, it can be approximated how the introduction of new (probiotic) species could lead to restoration of the stability and equilibrium of gut microbiota, but also when it could lead to its destruction. Regulation of the diversity and stability of the gut microbiota is therefore critical for the development of future therapies. The challenges for four areas of future therapy [obesity and metabolic syndrome, (recurrent infections), immune-related disorders, and the gut-brain axis] derived from the microbiome which were discussed during the workshop are summarized below.

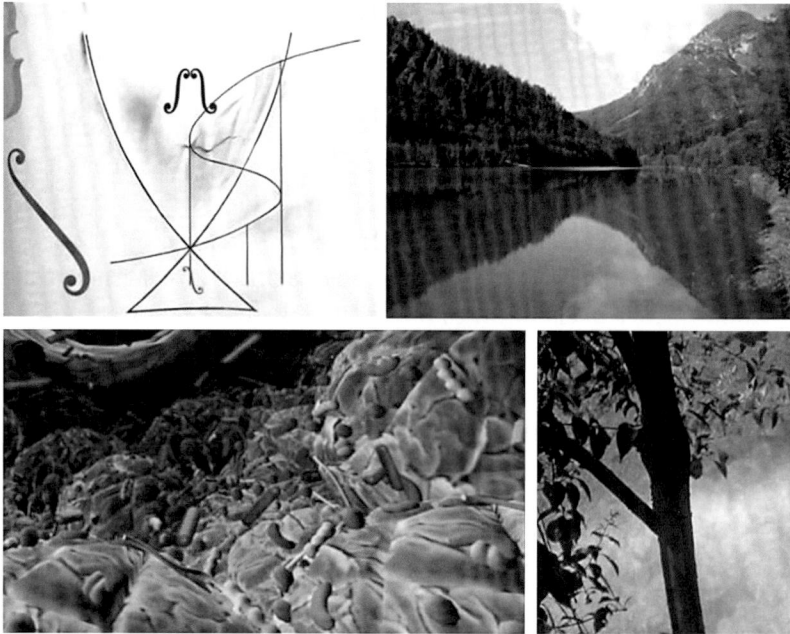

Fig. 1. Regulatory mechanisms for complex ecosystems, including lakes (upper right), rainforests (lower right), and gut microbiota (lower left). The upper left panel depicts Dali's interpretation of the mechanisms governing catastrophes. The upper right photograph is courtesy of Mag. A. Frauwall-ner, Graz, Austria; the lower left is courtesy of Dr. J. Doré, INRA, Micalis Institute, Jouy-en-Josas, France; and the lower right is courtesy of R. Lievendag, Middelburg, The Netherlands.

Obesity and Metabolic Syndrome

Obesity is threatening the world; it is becoming one of the most serious health problems of the 21st century, with increasing prevalence in both adults and children, and is one of the leading causes of death. Obesity is associated with the metabolic syndrome, which can lead to type 2 diabetes. Both authorities and scientists are looking for ways to prevent people from becoming obese, but also to prevent the incidence of health problems related to obesity, such as type 2 diabetes, which is irreversible. Host and environmental factors that affect the energy balance are major determinants. During the past 5–10 years, it has become clear that the human microbiota may play a role as well.

It started with the work of the group of Jeffrey Gordon [7] at Washington University, who showed an association between obesity and changes in the relative abundance of two dominant bacterial divisions in the microbiota: Firmicutes and Bacteroidetes. In obese mice, the ratio between Firmicutes and Bacteroidetes was in favor of Firmicutes, whereas the microbiota of the lean mice harbored relatively more Bacteroidetes. The same association was found in the microbiota of obese and lean human individuals. More strikingly, transplantation of an 'obese microbiota' into germ-free mice led to a significantly higher increase in body fat than colonization with a 'lean microbiota' [8]. Other studies showed a shift in the microbiota when obese individuals lost weight: the relative abundance of Firmicutes decreased, whereas the amount of Bacteroidetes increased and the ratio became more similar to that of lean individuals. Although these studies show a strong association between obesity and the microbiota composition, to date no European study has confirmed these results [9].

The transplantation of feces was recently executed in the FATLOSE trial at the Amsterdam Medical Center AMC, where obese men were transplanted with lean donor feces. The results showed a significant increase in peripheral insulin sensitivity after the transplantation [10]. Fecal transplant may not be the ideal solution in the end, but it is a first step towards microbiota management that leads to improvement in clinical symptoms.

Besides microbiota composition, the diversity or richness also seems to be relevant in health and health-related problems. Evidence is accumulating that a more diverse microbiota is related to health, whereas in disease, the diversity often seems to be decreased [1].

Rather than overall diversity, the presence or absence of specific species is associated with health and disease. *Akkermansia muciniphila*, *Faecalibacterium prausnitzii*, and *Bacteroides vulgatus* are among the species being extensively studied for potential associations with health and disease. When strong associations can be determined between the abundance of such species and health-related problems, these microbes can become biomarkers.

As indicated above, individuals differ in gut microbiota species composition, enterotype, and gene count. Recent research has shown that obese patients with a low gene count will be less susceptible to a low caloric diet than those with a high gene count. Furthermore, a higher inflammatory tone was found in association with a low gene count as well as elevated markers of risk of comorbidities [11, 12].

Diversity or gene count and the presence or absence of specific microbial species can become biomarkers and thus tools in the diagnosis and prognosis of obese patients. In addition, these markers can be relevant in personalized strategies for nutrition and medicine, for microbiota modulation, and perhaps in the prediction of disease risk in healthy persons. Probiotics, prebiotics, and synbiotics alone will never be sufficient to restore the disequilibrium between energy intake and energy expenditure [13]. However, as microbiota management tools, they can become a part of the total fight against obesity and related health problems in the near future.

(Recurrent) Infections

Worldwide, hundreds of millions of individuals are suffering from infections, often recurrent in nature, caused by bacteria, viruses, and parasites. These infections include *Clostridium difficile* infections (CDI), urinary tract infections, bacterial vaginosis, upper respiratory tract infections, malaria, and HIV.

With microbiomics it is possible to collect enormous amounts of genetic and biological data of microbiota, including benign and pathogenic microorganisms and their host, relatively cheaply and rapidly. With these data it will progressively be possible to develop rapid and cheap diagnostic techniques that will allow for more dedicated therapeutic interventions [14]. In turn these dedicated, and perhaps personalized, therapies will also be largely derived from microbiomics and will be directed towards re-establishing a healthy microbial community [15].

The microbial community in the gut is a dynamic entity that ideally protects us from infection by its so-called colonization resistance. A good example is the predisposition of patients to develop CDI upon treatment with antibiotics that can disrupt and destabilize the normal bowel microbiota. CDI is increasing, apparently due to the frequent use of antibiotics and the resulting extensive damage to the indigenous microbiota. Transplantation of feces from a healthy donor, so-called fecal bacteriotherapy, has been used successfully in severe cases of recurrent CDI, but there are risks of transmitting pathogens with this therapy [16]. Microbiomics could be employed successfully to sort out, perhaps even on the level of individual patients, which combination of gut-dwelling bacteria should be selected for a dedicated and safe probiotic therapy to cure recurrent CDI. Similar approaches could be followed to cure some of the other recurrent infections, such as vaginosis and urinary tract infections.

A good example was presented in a recent paper by Abreu et al. [17], in which they showed comparative microbiome profiling that chronic rhinosinusitis was caused by *Corynebacterium tuberculostearicum*, a common skin inhabitant. A probiotic strain of *Lactobacillus sakei*, isolated from the sinus resident microbiota, when applied as a nose spray proved to be effective in preventing sinusitis in mice. In this case, microbiomics was used to generate genetic and biological information leading to identification of a very specific bacterial strain as a candidate for therapeutic intervention.

In summary, microbiomics are the driving force behind novel diagnostic and bacteriotherapeutic methods that in the future will allow us to combat an array of infectious diseases on the level of the individual patient.

Immune-Related Disorders

Immune-related disorders encompass a broad category of diseases ranging from those which are characterized by an overactive Th2 system (allergies) to Th1-dominated autoimmune diseases. The hallmark of inflammatory bowel diseases such as ulcerative colitis and Crohn's disease is the ongoing mucosal inflammation. In his now classic paper, Jean Francois Bach [18] showed 10 years ago how immune-related disorders have increased over half a century. This trend has continued over the past decade [19].

In virtually all immune-mediated disease, an abnormal composition of gut microbiota has been found. However, whether this is cause or consequence is hard to establish. In a number of cases, such as allergy, the abnormal development of gut micro-biota precedes clinical onset of disease. The outcome of probiotic intervention in a number of immune-mediated diseases is discussed elsewhere in this book. In a number of instances, clinical success has been obtained in prevention and even management of immune-mediated diseases. However, variable results have also been obtained even in nearly identically designed studies [20, 21]. These different outcomes have been attributed to genetic differences in the host, differences in the environment, or in composition of gut microbiota. Observations by Lebeer et al. [22] indicated that differences can exist between different formulations of the same probiotic strain, in this case *Lactobacillus rhamnosus* GG; differential centrifugation may lead to loss of surface pili, the structures which are essential for adhesion to mucous surfaces.

An alternative way for microbiota management is the use of prebiotics. Prebiotics are currently being developed which mimic the size, linkage, and partly the building blocks and prebiotic functions of human milk oligosaccharides. Prebiotics also have been demonstrated to have direct or indirect immunomodulatory effects. Direct interaction of prebiotics with cells of the mucosal immune system can take place via lectins and lectin-like receptors. Galectins are a category of soluble-type lectins that may bind galactose/β-glycoside-containing glycans. Intestinal epithelial cells express galectins 2, 3, 4, 6, 7, and 9 abundantly in vivo. Emerging evidence indicates that galectins (intracellular or secreted) are regulators of immune homeostasis and inflammation: they facilitate cell-cell/matrix adhesion, induce T cell apoptosis, and promote chemotaxis. Administration of *Bifidobacterium breve* (TLR9-inducing) in combination with nondigestible oligosaccharides (galectin 9-inducing) reduces risk factors for asthma and respiratory allergy in infants [23].

The parallel development of the gut microbiome and the mucosal immune system during the first weeks and months of life [24, 25] offers a window of opportunity for intervention early in life. Primary prevention of allergic diseases by neonatal administration of probiotics and/or prebiotics indeed has been shown to be effective in a number of studies. The challenge for the future will be to target probiotics for existing immune-mediated disorders. For that, more insight into mechanisms and molecules, disease heterogeneity, other microbiota (skin [26], lung [27], oral, genital [28]), and risk/benefits will be needed.

Gut-Brain Axis

Functional gastrointestinal disorders (functional bowel disorder) are defined by symptom-based diagnostic criteria attributed to the gastrointestinal tract in the absence of pathologically based disorders [29]. There is a lot of published data showing perturbed microbiota composition in functional bowel disorder, particularly in irritable bowel syndrome (IBS). The general conclusion is that changes in the microbiota may contribute to symptoms in functional bowel disorder [30]. Although the exact etiology and pathophysiology of IBS is still unknown, many hypotheses/mechanisms which play a role have been put forward: dysregulation of the brain-gut axis and autonomic nervous system, visceral hypersensitivity, alterations in gut microbiota, altered levels of gastrointestinal neuropeptides and hormones, abnormal gastrointestinal motility, environmental and psychological factors (stress), and low-grade intestinal inflammation possibly related to alterations in gut microbiota [31].

A bidirectional neurohumoral communication system, known as the gut-brain axis, integrates the host gut and brain activities [32]. The intestinal microbiota communicates with the brain via this axis to influence brain development and behavior. This might influence a broad spectrum of diseases, including IBS, psychiatric disorders, and demyelinating conditions such as multiple sclerosis [33]. Putative mechanisms by which bacteria access the brain and influence behavior include bacterial products that gain access to the brain via cytokine release from the mucosal immune cells, release of gut hormones such as 5-HT from endocrine cells, or afferent neural pathways including the vagus nerve. Stress and emotions can also influence the microbial composition of the gut through the release of stress hormones or sympathetic neurotransmitters that influence gut physiology and alter the habitat of the microbiota. Alternatively, host stress hormones such as noradrenalin might influence bacterial gene expression or signaling between bacteria, and this might change the microbial composition and activity of the microbiota [33].

In the near future, research will be focused on the causal relationship between gut microbiota composition and the behavioral phenotype in animal and human studies. Personalized phenotypic characterization of the aberrations along the gut-brain axis will need to be investigated. Probiotics and prebiotics can be used in the treatment of IBS, but many questions remain (what strains, who can qualify, how long should treatment last, etc.). Understanding the mechanisms of microbiota modulation will be crucial, and modulation of barrier function is one of these. The effects of bacterial and host metabolites (tryptophan metabolites, fermentation products such as propionic acid, serine proteases) on the function of the gut should be elucidated, which also includes production of neurotransmitters by bacteria (GABA, noradrenaline, dopamine, acetylcholine, and 5-HT). Fecal transplant methods might be interesting to investigate the pre- and posttransplant behavioral changes in animal and even human studies.

Because stress plays an essential role in IBS, future research should also be focused on stress mechanisms: direct effects of microbiota on the HPA-axis, and 'dysbiosis' which could have both direct enteric nervous system as well as central nervous system effects. Attention should also be directed towards mast cells, which play an important role in neural pathways and immunity as well as in gut barrier function.

It can be concluded that more translation to human (clinical) research will be needed. More focus should be put on the small bowel with respect to microbiota composition and small bowel mucosa interactions. Furthermore, whether live or dead microbes should be used for this purpose remains to be determined. As in other cases of gut microbiota management, there is a need for relevant biomarkers and surrogate markers. Especially for intervention in the gut-brain axis, risk determinants and safety issues, e.g. in children, are of importance.

Conclusions and Future Outlook

Probiotics are defined as live microorganisms and therefore dead bacteria or bacterial products cannot be termed probiotic. Killed probiotic bacteria as well as secreted or purified active components of probiotic bacteria can, however, exert beneficial biological effects, as has been demonstrated in a number of in vitro studies (e.g. [34, 35]). In mice, the immunomodulating effect of *Bacteroides fragilis* can be mimicked by oral administration of the active component (PSA) in purified form [36]. Apparently, PSA when administered orally can reach the relevant niches within the intestine and interact with target cells. Such an approach would also be desirable for other microorganisms which cannot be cultured, e.g. *F. prausnitzii*. The prerequisite for successful application is identification and subsequent production of the active bacterial component. The advantages of the use of dead bacteria or bacterial components would include an even better safety profile and prolonged shelf life.

The interest in probiotics, and gut microbiota in general, by the scientific community has virtually exploded during the last decade. This is evidenced by the establishment of large research consortia, an overwhelming amount of publications in the highest-ranking journals, and data leading to new insights and possibilities for treatment of hitherto poorly understood diseases. The interest of the public at large for this subject can be seen in the Google Ngram viewer. This tool allows an analysis of the word count of all of Google Books. The data presented in figure 2 clearly show the exponential growth in the use of the word 'probiotics' over the last two decades. On the other hand, the scaling of the y-axis teaches us to be humble: current use of 'probiotics' is 1:5 million printed words (for comparison: 'love' is used 1: 5,000, 'hate' 1: 50,000).

Major challenges lie ahead, such as gaining a true and deeper understanding of the microbiota, and a better matching of probiotic mechanisms with clinical indication. Further understanding of the mechanisms which govern the stability of the microbial ecosystem will allow to design or select the most suitable strains of probiotics for a given condition and a given person. Until the factors which determine success or failure are known, it will be not be possible to implement a rational design of personal probiotic management. The most potent probiotic strains may still await identification and characterization. In a recent study on mice in bacteriotherapy of recurrent *C. difficile* disease, it was shown that the best combination of six strains, each ineffective on its own, included three previously unidentified species [37]. Also the emerging data on comparison of gut microbiota of inflammatory bowel syndrome patients with that of healthy individuals may lead to the identification of new bacterial strains with

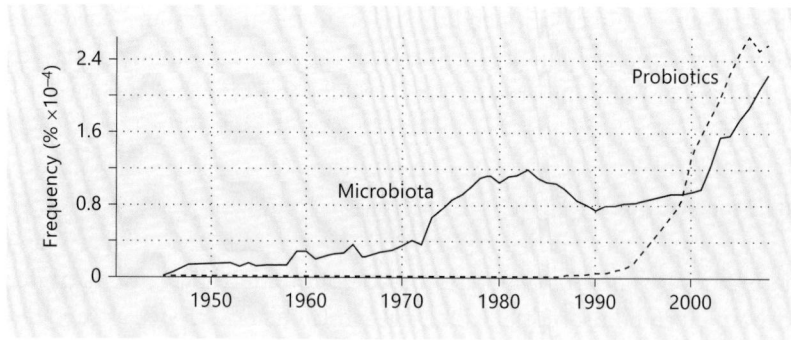

Fig. 2. The use of the terms microbiota and probiotics in the literature as assessed by Google Ngram viewer. The period covered is 1940–2008.

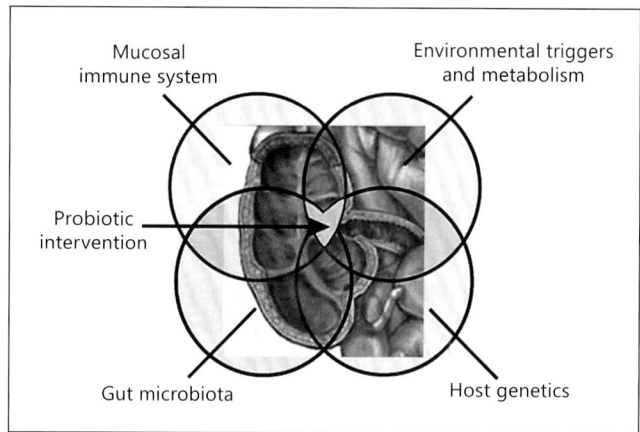

Fig. 3. Variables contributing to personalized gut microbiota management.

potential health benefits. Pivotal for any use of probiotics is the health benefit. A major target for probiotics therefore are the healthy at increased risk, whether because of lifestyle, age, genetic, or environmental influences (fig. 3). During every stage of research leading to new possibilities of (personalized) gut microbiota management, society should be informed of the risks and benefits, and unreasonable expectations should be managed. The road to personalized gut microbiota management will be long and winding, but the destiny makes it worthwhile.

Acknowledgement

We thank the moderators of the workshops on Obesity and Metabolic Syndrome (Nathalie M. Delzenne and Joel Doré), (Recurrent) Infections (Gregor Reid and Seppo Salminen), Immune-Related Disorders (Fergus Shanahan and Johan Garssen) and Gut-Brain Axis (Robert-Jan Brummer and John Bienestock) for their commitment, input, and support.

Rijkers · Mulder · Rombouts · Akkermans

References

1 Qin J, Li R, Raes J, Arumugam M, Burgdorf KS, Manichanh C, Nielsen T, Pons N, Levenez F, Yamada T, Mende DR, Li J, Xu J, Li S, Li D, Cao J, Wang B, Liang H, Zheng H, Xie Y, Tap J, Lepage P, Bertalan M, Batto JM, Hansen T, Le Paslier D, Linneberg A, Nielsen HB, Pelletier E, Renault P, Sicheritz-Ponten T, Turner K, Zhu H, Yu C, Li S, Jian M, Zhou Y, Li Y, Zhang X, Li S, Qin N, Yang H, Wang J, Brunak S, Doré J, Guarner F, Kristiansen K, Pedersen O, Parkhill J, Weissenbach J, MetaHIT Consortium, Bork P, Ehrlich SD, Wang J: A human gut microbial gene catalogue established by metagenomic sequencing. Nature 2010;464:59–65.

2 Wu GD, Chen J, Hoffmann C, Bittinger K, Chen YY, Keilbaugh SA, Bewtra M, Knights D, Walters WA, Knight R, Sinha R, Gilroy E, Gupta K, Baldassano R, Nessel L, Li H, Bushman FD, Lewis JD: Linking long-term dietary patterns with gut microbial enterotypes. Science 2011;334:105–108.

3 Arumugam M, Raes J, Pelletier E, Le Paslier D, Yamada T, Mende DR, Fernandes GR, Tap J, Bruls T, Batto JM, Bertalan M, Borruel N, Casellas F, Fernandez L, Gautier L, Hansen T, Hattori M, Hayashi T, Kleerebezem M, Kurokawa K, Leclerc M, Levenez F, Manichanh C, Nielsen HB, Nielsen T, Pons N, Poulain J, Qin J, Sicheritz-Ponten T, Tims S, Torrents D, Ugarte E, Zoetendal EG, Wang J, Guarner F, Pedersen O, de Vos WM, Brunak S, Doré J, MetaHIT Consortium, Antolín M, Artiguenave F, Blottiere HM, Almeida M, Brechot C, Cara C, Chervaux C, Cultrone A, Delorme C, Denariaz G, Dervyn R, Foerstner KU, Friss C, van de Guchte M, Guedon E, Haimet F, Huber W, van Hylckama-Vlieg J, Jamet A, Juste C, Kaci G, Knol J, Lakhdari O, Ayec S, Le Roux K, Maguin E, Mérieux A, Melo Minardi R, M'rini C, Muller J, Oozeer R, Parkhill J, Renault P, Rescigno M, Sanchez N, Sunagawa S, Torrejon A, Turner K, Vandemeulebrouck G, Varela E, Winogradsky Y, Zeller G, Weissenbach J, Ehrlich SD, Bork P: Enterotypes of the human gut microbiome. Nature 2011; 473:174–180.

4 Dakos V, Carpenter SR, Brock WA, Ellison AM, Guttal V, Ives AR, Kéfi S, Livina V, Seekell DA, van Nes EH, Scheffer M: Methods for detecting early warnings of critical transitions in time series illustrated using simulated ecological data. PLoS One 2012;7:e41010.

5 Lenton TM, Livina VN, Dakos V, van Nes EH, Scheffer M: Early warning of climate tipping points from critical slowing down: comparing methods to improve robustness. Philos Transact A Math Phys Eng Sci 2012;370:1185–1204.

6 Veraart AJ, Faassen EJ, Dakos V, van Nes EH, Lürling M, Scheffer M: Recovery rates reflect distance to a tipping point in a living system. Nature 2011;481: 357–359.

7 Ley RE, Turnbaugh PJ, Klein S, Gordon JI: Microbial ecology: human gut microbes associated with obesity. Nature 2006;444:1022–1023.

8 Turnbaugh PJ, Ley RE, Mahowald MA, Magrini V, Mardis ER, Gordon JI: An obesity-associated gut microbiome with increased capacity for energy harvest. Nature 2006;444:1027–1031.

9 Tremaroli V, Bäckhed F: Functional interactions between the gut microbiota and host metabolism. Nature 2012;489:242–249.

10 Vrieze A, Van Nood E, Holleman F, Salojärvi J, Kootte RS, Bartelsman JF, Dallinga-Thie GM, Ackermans MT, Serlie MJ, Oozeer R, Derrien M, Druesne A, Van Hylckama Vlieg JE, Bloks VW, Groen AK, Heilig HG, Zoetendal EG, Stroes ES, de Vos WM, Hoekstra JB, Nieuwdorp M: Transfer of intestinal microbiota from lean donors increases insulin sensitivity in individuals with metabolic syndrome. Gastroenterology 2012;143:913–916.

11 Flint HJ: Obesity and the gut microbiota. J Clin Gastroenterol 2011;45(suppl):S128–S132.

12 Larsen N, Vogensen FK, van den Berg FW, Nielsen DS, Andreasen AS, Pedersen BK, Al-Soud WA, Sørensen SJ, Hansen LH, Jakobsen M: Gut microbiota in human adults with type 2 diabetes differs from non-diabetic adults. PLoS One 2010;5:e9085.

13 Nicholson JK, Holmes E, Kinross J, Burcelin R, Gibson G, Jia W, Pettersson S: Host-gut microbiota metabolic interactions. Science 2012;336:1262–1267.

14 Budding AE, Grasman ME, Lin F, Bogaards JA, Soeltan-Kaersenhout DJ, Vandenbroucke-Grauls CM, van Bodegraven AA, Savelkoul PH: IS-pro: high-throughput molecular fingerprinting of the intestinal microbiota. FASEB J 2010;24:4556–4564.

15 Lozupone CA, Stombaugh JI, Gordon JI, Jansson JK, Knight R: Diversity, stability and resilience of the human gut microbiota. Nature 2012;489:220–230.

16 Postigo R, Kim JH: Colonoscopic versus nasogastric fecal transplantation for the treatment of Clostridium difficile infection: a review and pooled analysis. Infection 2012;40:643–648.

17 Abreu AN, Nagalingam NA, Song Y, Roediger FC, Pletcher SD, Goldberg AN, Lynch SV: Sinus microbiome diversity depletion and Corynebacterium tuberculostearicum enrichment mediates rhinosinusitis. Sci Transl Med 2012;4:151ra124.

18 Bach JF: The effect of infections on susceptibility to autoimmune and allergic diseases. N Engl J Med 2002;347:911–920.

19 Palmer DJ, Metcalfe J, Prescott SL: Preventing disease in the 21st century: the importance of maternal and early infant diet and nutrition. J Allergy Clin Immunol 2012;130:733–734.

20 Kalliomäki M, Salminen S, Arvilommi H, Kero P, Koskinen P, Isolauri E: Probiotics in primary prevention of atopic disease: a randomised placebo-controlled trial. Lancet 2001;357:1076–1079.

21 Kopp MV, Goldstein M, Dietschek A, Sofke J, Heinzmann A, Urbanek R: Lactobacillus GG has in vitro effects on enhanced interleukin-10 and interferon-gamma release of mononuclear cells but no in vivo effects in supplemented mothers and their neonates. Clin Exp Allergy 2008;38:602–610.

22 Lebeer S, Claes I, Tytgat HL, Verhoeven TL, Marien E, von Ossowski I, Reunanen J, Palva A, Vos WM, Keersmaecker SC, Vanderleyden J: Functional analysis of *Lactobacillus rhamnosus* GG pili in relation to adhesion and immunomodulatory interactions with intestinal epithelial cells. Appl Environ Microbiol 2012;78:185–193.

23 de Kivit S, Saeland E, Kraneveld AD, van de Kant HJ, Schouten B, van Esch BC, Knol J, Sprikkelman AB, van der Aa LB, Knippels LM, Garssen J, van Kooyk Y, Willemsen LE: Galectin-9 induced by dietary synbiotics is involved in suppression of allergic symptoms in mice and humans. Allergy 2012;67:343–352.

24 Hooper LV, Littman DR, Macpherson AJ: Interactions between the microbiota and the immune system. Science 2012;336:1268–1273.

25 Maynard CL, Elson CO, Hatton RD, Weaver CT: Reciprocal interactions of the intestinal microbiota and immune system. Nature 2012;489:231–241.

26 Naik S, Bouladoux N, Wilhelm C, Molloy MJ, Salcedo R, Kastenmuller W, Deming C, Quinones M, Koo L, Conlan S, Spencer S, Hall JA, Dzutsev A, Kong H, Campbell DJ, Trinchieri G, Segre JA, Belkaid Y: Compartmentalized control of skin immunity by resident commensals. Science 2012;337:1115–1119.

27 Blainey PC, Milla CE, Cornfield DN, Quake SR: Quantitative analysis of the human airway microbial ecology reveals a pervasive signature for cystic fibrosis. Sci Transl Med 2012;4:153ra130.

28 Reid G: Probiotic and prebiotic applications for vaginal health. J AOAC Int 2012;95:31–34.

29 Drossman DA: The functional gastrointestinal disorders and the Rome III process. Gastroenterology 2006;130:1377–1390.

30 Simren M, Barbera G, Flint HJ, Spiegel BMR, Spiller RC, Vanner S, Verdu EF, Whorwell PJ, Zoetendal EG: Intestinal microbiota in functional bowel disorders: a Rome foundation report. Gut 2013;62:159–176.

31 Ghoshal UG, Ratnaker S, Ghoshal U, Gwee K-A, Ng SC, Quigley MM: The gut microbiota and irritable bowel syndrome: friend or foe? Int J Inflam 2012;2012:151085.

32 Forsythe P, Kunze WA, Bienenstock J: On communication between gut microbes and the brain. Curr Opin Gastroenterol 2012;28:557–562.

33 Collins SM, Surette M, Bercik P: The interplay between the intestinal microbiota and the brain. Nature Rev Microbiol 2012;10:735–742.

34 Orlando A, Refolo MG, Messa C, Amati L, Lavermicocca P, Guerra V, Russo F: Antiproliferative and proapoptotic effects of viable or heat-killed *Lactobacillus paracasei* IMPC2.1 and *Lactobacillus rhamnosus* GG in HGC-27 gastric and DLD-1 colon cell lines. Nutr Cancer 2012;64:1103–1111.

35 Fujiki T, Hirose Y, Yamamoto Y, Murosaki S: Enhanced immunomodulatory activity and stability in simulated digestive juices of *Lactobacillus plantarum* L-137 by heat treatment. Biosci Biotechnol Biochem 2012;76:918–922.

36 Surana NK, Kasper DL: The yin yang of bacterial polysaccharides: lessons learned from *B. fragilis* PSA. Immunol Rev 2012;245:13–26.

37 Lawley TD, Clare S, Walker AW, Stares MD, Connor TR, Raisen C, Goulding D, Rad R, Schreiber F, Brandt C, Deakin LJ, Pickard DJ, Duncan SH, Flint HJ, Clark TG, Parkhill J, Dougan G: Targeted restoration of the intestinal microbiota with a simple, defined bacteriotherapy resolves relapsing *Clostridium difficile* disease in mice. PLoS Pathog 2012;8:e1002995.

Ger T. Rijkers
Department of Sciences, Roosevelt Academy
Lange Noordstraat 1, PO Box 94
NL–4330 AB Middelburg (The Netherlands)
E-Mail g.rijkers@roac.nl

Author Index

Akkermans, L.M.A. 186

Barrett, E. 56
Berni Canani, R. 128
Brandtzaeg, P. 43
Brigidi, P. 25
Bruzzese, E. 139
Buccigrossi, V. 9

Collado, M.C. 95
Cosenza, L. 128
Cryan, J.F. 56

Di Costanzo, M. 128
Dinan, T.G. 56

Fitzgerald, G.F. 56, 103
Forssten, S.D. 171

Granata, V. 128
Guandalini, S. 151
Guarino, A. XI, 9
Guarner, F. 17

Hill, C. 178
Hojsak, I. 161

Ibrahim, F. 171
Indrio, F. 79
Isolauri, E. 95

Kobayashi, K.S. 32

Laitinen, K. 95
Leone, L. 128
Lo Vecchio, A. 139

Luoto, R. 95
Lynch, S.V. 64

Morelli, L. 1
Mulder, L. 186
Murphy, E.F. 103

Neu, J. 122
Nicastro, E. 9
Nocerino, R. 128

O'Toole, P.W. 25, 56, 103
Ouwehand, A.C. 171

Patton, T.J. 151
Pezzella, V. 128
Pigneur, B. 72
Power, S.E. 103

Quigley, E.M.M. XI, 56, 87, 103

Riezzo, G. 79
Rijkers, G.T. 186
Robles Alonso, V. 17
Rombouts, F.M. 186
Ross, R.P. 56, 103
Röytiö, H. 171
Ruberto, E. 139
Ruemmele, F.M. 72

Salminen, S. 95
Shamir, R. 161
Shanahan, F. 56
Stanton, C. 56, 103

Walker, W.A. XI

Subject Index

intestinal epithelium
    antibody translocation 47–49
    innate immunity ontogeny 50, 51
    microbiota crosstalk 51, 52
  microbe sources 65
  regulatory T cell peripheral
    induction 52
infection resistance studies, *see* Infection
metagenomics 17–19, 186
NOD-like receptor regulation 36–38
obesity and metabolic syndrome studies
  adults
    animal studies 104
    cardiovascular disease studies 109,
      110
    diabetes type 2 studies 108, 109
    human studies 105
    mechanisms of obesity linkage 105–
      108
    nonalcoholic fatty liver disease
      studies 110
  children and infants 98, 99
RegIIIγ regulation 34, 35
species
  adult enterotypes 22, 23
  developmental changes 11–13, 73
  identification techniques 19, 20, 130
Irritable bowel syndrome
  maternal separation model 58
  probiotic studies
    adults 89, 90
    children 83, 84
    dose-effect relationship 157
  overview 60, 61

Lactose intolerance, probiotic studies 173, 174
Listeria, commensal/probiotic bacteria
  competition with pathogens 181

Maternal separation, gut-brain axis effects 58,
  59
Metabolic syndrome, *see* Obesity
Metagenomics, intestinal microbiota 17–19
Microbe-associated molecular patterns 44, 52
Microbiota, *see* Intestinal microbiota

Necrotizing enterocolitis
  clinical presentation 122, 123
  laboratory features 123
  pathogenesis 123, 124
  prevention

    human milk 124, 125
    prospects for study 125, 126
  probiotic dose-effect relationship 156
  probiotic therapy 14, 125, 147
NOD-like receptor
  intestinal microbiota regulation 36–39
  NOD-like receptor protein 6 38, 39
Nonalcoholic fatty liver disease, intestinal
  microbiota studies 110

Obesity
  adult obesity and metabolic syndrome
    intestinal microbiota
      animal studies 104
      cardiovascular disease studies 109,
        110
      diabetes type 2 studies 108, 109
      human studies 105
      mechanisms of obesity linkage 105–
        108
      nonalcoholic fatty liver disease
        studies 110
    probiotic studies
      animal studies 111–114
      human studies 114–117, 188,
        189
      overview 110, 111
  children and infants
    intestinal microbiota studies 98, 99
    probiotic reshaping of early intestinal
      microbiota 100, 101
  comorbidity 103
  energy imbalance 95, 96
  feces transplantation studies 189

Prebiotics, intestinal microflora effects 191
Probiotics
  age-related functional feeding
    adults 173, 174
    children 172, 173
    elderly 175
  airway disease studies 67, 68
  γ-aminobutyric acid receptor expression
    effects in brain 60, 61
  definition
    country-specific definitions 6
    FAO/WHO 2001 document 3, 4, 151
    FAO/WHO 2002 guidelines 4–6
    overview 1–3
  dose-effect relationship
    adverse effects of high dose 157, 158